# Cosmos and Metacosmos

# COSMOS AND METACOSMOS

ROBERT WESSON

La Salle, Illinois

First printing 1989



Printed and bound in the United States of America.

**Library of Congress Cataloging-in-Publication Data**
Wesson, Robert G.
Cosmos and metacosmos / Robert Wesson.
p. cm.
Bibliography: p.
Includes index.
ISBN 0-8126-9091-5. — ISBN 0-8126-9092-3 (pbk.)
1. Cosmology. 2. Evolution. 3. Civilization. 4. Mind and body.
5. Ethics. I. Title
BD511.W3 1989
113—dc20

89-8709
CIP

# Table of contents

# Acknowledgements

This work owes much to many persons with whom the author has discussed various ideas, as well as many writers, whose works are cited. Patricia de la Torre, Jessica Syme, Heinz Pagels, and Menahem Schiffer have kindly read more or less of the manuscript, and they have made valuable suggestions. Sarah Blair, Jennifer Bravinder, Margaret Smith, and Carol Wesson deserve credit for many textual improvements. Most of all, however, the author wishes to express appreciation to William Warren Bartley III for his intelligent critique and encouragement.

# Preface

Conditions of living have been altered more in the past hundred years than in the previous thousand. Change accelerates dizzily; the informational storehouses of humanity are doubled every decade or so, and their content must have been multiplied at least ten times over within the memory of the middle-aged. Yet our ways of seeing ourselves in relation to our universe have changed much less than our knowledge of it. Modern science has not much helped philosophy to give a sounder orientation to contemporary society.

Science instead contributes to confusion. One can learn indefinitely about a thousand things, from the biology of bacteria to the effect of drugs on the brain and the caprices of electrons, but deeper answers are missing. As one reaches the outer limits of knowledge, the mind is ever more adrift, and the conviction of many scientists that everything must be reducible to material entities muddies the intellectual waters.

What is one to think of the fact that contemporary physics finds ever more complications in the smallest and supposedly simplest things that can be registered? Why are scientists driven to postulating wholly unobservable entities, such as the multidimensional coils, infinitely tinier than a proton, the vibrations of which are called upon to make the qualities of particles? Or of the outlandish "strings" of inconceivable mass, invented to put galaxies in their places? Or of the fact that physicists take seriously the bizarre hypothesis that the universe replicates itself an indefinite number of times with each quantum event, so that in any instant one universe would become a near infinity of sister-universes, each of which would go into a near-infinity of branches in the next instant, and so forth?

Biology suffers a different malaise. Despite an enormous accumulation of facts, biologists have advanced little beyond the premodern ideas of Darwin (whose genius should not be demeaned) in the core question of how and why plants and animals have taken on marvellous forms and ways for which natural selection offers no explanation. The emergence of human intelligence is little less mysterious than it was before the discovery of the first hominoid fossils in the latter part of the 19th century. Psychology has amassed a huge volume of information about the brain and mind, but it has contributed little to our understanding of ourselves, and the healing of mental diseases has hardly progressed for generations.

Our view of ourselves in relation to others and the universe should be based on understanding reality, yet the abundance of knowledge has brought little enlightenment. The present work seeks to reduce this anomaly by bringing together fundamental results of modern scientific inquiry, drawing them as far as possible into a single framework and asking what they imply about the meaning of human existence.

For this, it is necessary to survey a wide range of scientific concepts. If we are to find any meaning inherent in the cosmos, it must be of the entirety; deeper comprehension requires the synthesis of all major aspects. What we learn of the mind is quite as relevant as what we learn of the electron. Hence this work touches, in successive chapters, on fundamentals of physics, biology, history, and psychology, and in a final chapter draws some moral consequences. Modern science demands specialization and looks askance on the superficiality of excessive breadth; this book, however, to fulfill its purpose, has to be unconventionally broad.

Yet it is not excessively difficult to arrive at a broad vision by surveying the chief aspects of our complex being. One can appreciate the thrust and principal conclusions of modern physics without mastering higher order differential equations and grasp the essentials of what is known of the mind without delving into details of neurology. The fundamentals of modern astronomy, biology, and psychology are at least as accessible as a Joycean novel and require no more concentration than a good chess game.

The nonspecialist, moreover, has advantages of perspective over the professional engrossed in a particular specialty.

The starting point of this book is the seemingly well-established fact that the material universe had a beginning. This obviously implies that it arose from something, a matrix with the ability not only to engender matter but to generate complexity and order. Our reality consists of material substance, born in the original super-fireball of the super-explosion commonly called the Big Bang, coming out of and expanding within an order-giving matrix. This generative matrix is here called the "metacosmos," "meta" meaning transcending, as in "metaphysics," the study of being, or "metamathematics," the study of the nature of mathematics. Although we cannot hope fully to understand the metacosmos, we can know something of it because we are its creatures, share something of its being, and perceive its realization in the material universe.

Thinking in terms of cosmos and metacosmos makes many things less incomprehensible, from the inordinate complexity of the physical universe to the astonishing inventiveness of living nature and the ambiguities of the relations of the human mind to its physical basis, as materiality is mingled with spirituality, or bestiality is alloyed with divinity. Complexly organized beings unfold ever greater creative powers as the cosmos, the orderly universe, fulfills the order-making metacosmos.

The result of this line of thought, as laid out in the following pages, is a nontheistic nonmaterialism. This lies midway between the two traditional fundamental ways of thinking about what are called "higher things": the belief that the world is the work of a great personality who watches over it and perhaps intervenes on occasion to set things right; and the theory that material particles are the totality of existence, for the reality and basic properties of which there can and need be no accounting. Both of these views are incomplete and unsatisfactory for those who are unable to take answers on faith but are impressed with the inadequacy of materialistic interpretations. The most reasonable approach seems to compromise between fideism and materialism: the material cosmos came out of and is profoundly related to a transcendant essence.

This broad scheme does not, of course, give specific answers. We can prove nothing about everything. It is no more possible to demonstrate the existence of the metacosmos than the existence or nonexistence of God; that it continues to shape the development of the universe is evidenced only indirectly. But it seems clear that "creation is really a continuing process,"[1] and it is reasonable to credit the continuation of the process to the same forces that set it in motion.

We cannot attribute to the metacosmos everything for which we do not find a material explanation, but the scheme of cosmos and metacosmos should help to order thoughts about profound questions. It seems sensible to regard potentialities of intelligence, or of creative ordering, as fundamental to the entirety of things rather than to postulate that creativity is mysteriously incorporated in the elementary particles, which somehow generated intelligence—the conventional position of science, which is presented as hardheadedly materialistic but which is essentially mystic. There is no reason to attribute to the metacosmos human moral qualities ordinarily ascribed to a divinity, such as benevolence and justice. The metacosmos cannot be conceived as personal; "personal" pertains to individuals of our mortal species. But it has clear implications for the conduct and purpose of human life. Contemplating the metacosmos, one sees reality as a fertile mingling of matter and pattern wherein designs grow ever more complex as understanding expands.

The concept of the metacosmos rests not on any single fact but on the appreciation of a multifaceted reality, which this book can only briefly survey. The full significance of such ideas cannot be immediately apparent; if it were, the following pages would be superfluous.

# 1

# The material cosmos

## Evolving perspectives

For thousands of years, prophets and philosophers have meditated about the nature of their world and their place in it. From the Vedas, the Bible, and Confucius to Immanuel Kant and modern thinkers, many writers have left messages of enduring value, even for our times so changed and our problems so different from those they knew. Yet for modern thinking their validity is more moral than intellectual; they serve more for inspiration than analysis. Modern civilization, bustling with all manner of preoccupations, flooded by information and misinformation and overwhelmed by all manner of needs and opportunities, lacks clearly reasoned personal and social goals.

Understanding feeds on information, and we have at our disposal riches of information beyond the imagination of previous centuries. If we can even very partially take into consideration the new knowledge, we should be able to reach greater depth than the wisest of earlier times. Science has not ripped away the curtain hiding the mysteries of the universe, but it has slightly lifted the edges.

With the maturation of human inquiry into the world, many discoveries have profoundly affected our vision of our world, replacing near-to-hand, commonsense notions with deeper and better informed explanations. The stage on which we play out our lives seems obviously flat and covered by a sky-dome, across which the sun travels daily. The most advanced ancient Greek thinkers surmised that we live on a turning globe, but it was still posited to be the fixed center of things. Copernicus' teaching in the 16th century that this fair earth circles around a much larger sun was shocking and widely rejected, but it gradually came to be accepted in order to make sense of the movements of celestial bodies. In this century, we have been taught to regard the earth as a merest speck in an unimaginably immense void, special only in the accident of our living here.

Closely linked to the loss of centrality was the despiritualization of the world and its subjugation to fixed rules or natural law. The idea of impersonal and orderly cause and effect, which was pioneered by the ancient Greek philosophers, has grown rather steadily from the 13th century or earlier. After the 15th to 17th century voyages of discovery, learning about many different cultures brought into thinking a relativism that has not ceased to grow. A sign of the rise of rationalism was the renewed practice of dissection of the human frame in the 16th century after the lapse of anatomical studies since Roman times. In the 17th century, the analytical philosophy of René Descartes formalized the empirical approach. It leaped forward with Newton's contemporaneous discovery of laws of gravitation and mechanics, which governed the solar system and placed the planets and falling apples under the same rule. Newton went on to an unexampled burst of explanation as he calculated the tides, the earth's equatorial bulge, the precession of the equinoxes, the movement of comets, and changes in the moon's orbit.

This astounding achievement gave confidence that the world was to be understood in terms of mechanical cause and effect if adequately studied. Scientists happily thought they had learned that the world was a machine, a great clockwork in the metaphor of the most impressive apparatus of the day (like the computer in our time). Everything was to be comprehended and predicted in mechanical terms as the straightforward result of the laws of force and motion: in principle all phenomena were as predictable as the motions of the planets—a belief as attractive to the scientific mind as it was (and is) unrealistic. The emergence of chemistry from alchemy in the 18th century seemed to give more substance to the hope of subjecting the mysteries of being to simple regularities called laws of nature. The enormous success of the whole industrial-scientific revolution has been a grand confirmation of this materialistic approach.

Already in the 18th century, a century before Darwin, thinkers were coming to see the human species as part of the community of living creatures, which Carl Linnaeus put into systematic order. The discovery of obviously near-human African apes made a deep impression. In the 19th century, it became evident that the earth's age was to be counted not in the few millennia indicated by Scripture but in millions of years, or billions as later became known. This set the stage for the even more distressing thesis, brilliantly documented and elaborated by Charles Darwin, that humans do not stand isolated and apart as specially created beings but are only the present climax of a long chance-directed development—a thesis not only contrary to the obviously enormous differences between humans and animals but injurious to our self-esteem. By the end of the 19th century, not only had the world been subjected to fixed, basically mechanistic regularities built into the nature of things, but we ourselves had been downgraded from small images of God to improved apes, creatures whose proud intelligence was attributed to senseless chance, the workings of random variation sieved by natural selection.

The first decades of the 20th century further discomfited human pride. Sigmund Freud taught that the outwardly rational, civilized personality was held in thrall by a dark unconscious, driven mostly by lust, or libido. Marxism, developed in the 19th century and becoming influential in the 20th, regarded human society as motivated by economic and class interests, over which culture and political institutions lay as a cloak for exploitation. The personality was also demeaned by the psychology of behaviorism, preached by J.B. Watson and successors through B.F. Skinner. This approach virtually denied that anything to be called the "mind" really existed; instead, behaviorists thought in terms of bundles of conditioned reflexes.[2] A little later, in the 1940s and 1950s, long molecules of nucleic acid were found to carry the full set of directions for making a plant, an animal, or a human. Molecular biology reinvigorated the hope (as had Newton's laws three centuries earlier) that everything could be reduced to the interactions of material particles.

But physics meanwhile was dissolving certitudes. Investigation of radiation and the constituents of matter early in this century led to the quantum theory governing elementary particles; this entirely non-Newtonian mechanics replaced predictability with statistical probability and made the cosmos seem much less reliable. Einstein's Theory of Relativity had a similar effect. It abolished the traditional absolutes of time, length, and even mass, allowing only relative measurements; the one fixed and certain quantity was the velocity of light, something totally inaccessible to human senses. The naive understanding of the material world was thoroughly confounded.

In the last decades, science has continued piling up a stupendous volume of facts, acquiring fantastic tools of research, and giving detailed understanding of countless phenomena. At the beginning of this century eminent scientists still doubted whether atoms were real entities; now it is possible to photograph individual atoms and play with them. Half a century ago viruses were known only as mysterious infectious agents characterized by their ability to pass through a fine filter; now their molecular architecture can be detailed. Early in this century, it was not known that other agglomerations of stars like ours existed; now astronomy reaches out to photograph galaxies being formed when the universe was young, and cosmology has attained an impressive degree of understanding of the beginnings of all things. The store of scientific information about the material universe and ourselves has increased at least tenfold in the past fifty years.

Ironically, the marvellous expansion of knowledge and human capabilities has had a negative effect on the human self-image. Modern man has been induced to feel weaker, more ignoble, and more insignificant just as his mastery of nature was expanding fabulously. This is the result not only of specific knowledge or theories, such as Darwinian evolution, Freudian and behavioral psychology, and molecular biology, but of success in understanding nature in general. The overall triumph of science suggests that everything, including the human mind, is to be understood in materialistic-scientific terms. It seems to follow, that human life and thought, being ultimately a product of chance, has no possible meaning or purpose. In the ancient phrase, all is vanity.

This outlook claims the sophistication of scientific backing. In reality, however, it suffers naivete, just as it was naive of thinkers of the 18th century to suppose that the success of Newtonian mechanics proved that man was essentially a machine. It is contradictory that the far grander intellectual edifice of today seems to degrade its maker, that the fantastic achievements of science are taken to trivialize the makers of science and reduce them to complex aggregations of matter. Logically, science cannot diminish its creator, just as computers, no matter how capable of something like intelligence, cannot prove the incapability of their designers. The knower must be superior to the known.

A profounder world view must take the results of science into consideration without being overwhelmed and compressed by them. If science could indeed find a set of rules encompassing all reality, a grandly ordering simplification, this might imply that the mind must take its place in the finite sphere and can look to nothing beyond it. But science approaches no such simplification. It proceeds always to greater complexity, and each answer leads to new questions and new mysteries. Modern science has found not a closed but an intellectually open universe.

We should consequently seek a more realistic world view, a truer vision of our position in the entirety of being. Such a view must be based on modern scientific knowledge. We cannot rationally take refuge in faith or hope for mystic enlightenment, which is private to each meditator, but must look to broad inferences from what has been learned, primarily about the material universe. Science, indeed, is moving in this direction; some physicists, peering toward horizons of the unknowable, are much readier than chemists, biologists, or psychologists to perceive intelligent purpose in the nature of things.

## THE GENESIS

Many thinkers, preferring the regular and predictable, would like to have their universe existing forever more or less as is. It is a cherished and essential axiom of science that the laws of nature are everywhere and always the same; it assumes that if conditions are identical an experiment will turn out identically in New York and in Tokyo, and now as a century ago or a century hence. Science assumes—and the assumption seems well supported—that particles have always interacted and will always interact in the same way. By extension, it appeals to the sense of intellectual order to believe that the universe is spatially uniform, no direction and no region being favored or fundamentally different from any other, as the Copernican principle suggests. Further, there should be no preferred time; our age should be cosmologically like any other.

This view prevailed until recently. The static, steady-state universe is comforting because it eliminates the need to wrestle with the question of a beginning. It has simply been there forever. An always existing universe is equally agreeable from the materialistic viewpoint in that it can be conceived as being the whole of reality, containing all causation within itself and embodying no conceivable purpose. If there was no creation, there is no need to postulate a Creator. The universe remains fundamentally miraculous and inherently unaccountable, to be sure; but if it represents the entirety of existence, we may assume it to be wholly accessible to human understanding. The knowable is sharply set off from the unknowable.

For the scientific mind these virtues have seemed to outweigh the chief defect of the idea of a universe without beginning or end, the implication that nothing is really new, as all combinations must have been tried over and over in the infinite expanse of time; that intelligent life has grown up and been extinguished an infinite number of times, like the Hindu wheel of eternal recurrence, ever revolving and leaving no trace of its meaningless motion.

Pleasing or not, the idea of basic changelessness raises many problems, and no one has been able to draw up a convincing blueprint for a changeless condition of a very dynamic cosmos. One difficulty is that it is hard to devise a universe in equilibrium having a continual flux of energy, as ours has and as seems indispensable for lifelike processes. An infinitely old universe should be filled up with light; it would be difficult to account for the black background of the stars. The night sky is dark because there has not been nearly enough time for stars to illuminate it, and they will be consumed long before they can.[3] As matter is continually being burned or degraded by nuclear fusion in stellar interiors, a steady-state universe requires means of both producing new fuel (hydrogen) and getting rid of the ashes (heavier elements). Attempts have been made to solve this problem by hypothesizing the continuous generation of matter in empty space, leading to condensation into galaxies and stars. It was theorized that space did not fill up with light and matter because it expanded rapidly enough to maintain a constant density. However, this rather contrived theory did not fit observed facts.

Efforts to rationalize a universe without a beginning[4] have been abandoned as evidence has accumulated that our cosmos came into being between 10 and 20 billion years ago. Einstein's General Relativity (1917) pointed in the direction of an expanding universe, but the idea of a steady state was so generally accepted that Einstein was misled into introducing an arbitrary constant in his equations as a stabilizer.[5] In the late 1920s, the spectral red shift of distant galaxies was discovered. Edwin Hubble tried at first to interpret this phenomenon in terms of a static universe, but it soon became accepted as a Doppler effect, evidence that the galaxies are all rushing away from us and each other like the debris set flying by an explosion. This was confirmed by the discovery in 1965 of an all-pervasive uniform background radiation, the wavelength of which corresponds to the light that the original fireball would have emitted, stretched out by the expansion of space.

Physicists thereupon began to take seriously the idea, first proposed in 1927 by Georges Lemaître, of a beginning in a great explosion; and various confirmatory facts, such as the relative abundance of hydrogen, helium, and deuterium, fit well with the hypothesis of a Big Bang. The beginning of our universe perhaps 15 billion years ago seems as much a certainty as anything so long ago can be.

The scenario of cosmic beginnings rests on the reasonable assumption that the laws of nature are changeless. Scientists in effect run the motion picture of our universe backward; and despite difficulties in analyzing the very early course of events, they come to a remarkably detailed and coherent account of events from less than a trillionth of a second after the beginning, which was presumably at a singularity or dimensionless point. The chief basis for analysis of conditions at this early time, when temperatures were $10^{30}$ K or more, is the behavior of particles brought very nearly to the speed of light in accelerators. Before the first trillionth of a second (nanosecond) the energies were much too high to be matched in any apparatus, and speculation has a free field.

There are necessarily many uncertainties about the course of formation of the star-studded universe.[6] Most troublesome is how matter came to be distributed in galaxies, clusters of galaxies, and clusters of clusters, some 60 million light years across. For our purposes, however, the finiteness of past time itself is the relevant fact. Nothing can be said about the beginning of something that goes on forever; but if the entirety of space-time-matter had a beginning, it must have been generated out of something transcending our space-time. If it all came into being, or was made, at a certain point, there had to be causation from without, something anterior and exterior to the observable cosmos capable of giving rise to a very complex development.

"Anterior," however, does not mean earlier in our time-frame. We suppose that time, space, and matter, as we know them, began together, and the expansion of the universe gives the otherwise rather enigmatic direction of time. There was no "before" our universe in our time dimension (unless we believe there has been a recurrent cycle of expansion and contraction to expand again, for which there is no evidence). This may become more comprehensible if we recall that, as Einstein showed, time and space are not absolutes but are subject to distortion; and in black holes (the existence of which is believed to be certain) space and time lose their meaning. It is speculated that numerous universes may have come into existence together but flew off with their own time-frames, or that space-time systems were initially discontinuous and subject to breaking and joining.[7] But such speculations do not bear upon the central thesis of the formation of this cosmos from an independently existent womb of being. The material cosmos is not the totality of existence but a derivative of something primordial.

## Metacosmos

The generative matrix of the material cosmos must be envisioned as outside our space-time but having the capacity to create time and space, which define our material reality. In other words, our space-time can be regarded as embedded in an existence of higher dimensionality or its equivalent, somewhat as a 2-dimensional surface may be created by slicing a 3-dimensional object.[8] This idea comes easily to modern mathematics, which treats multidimensionality with abandon and freely postulates as many dimensions as seem useful to a physical or mathematical scheme. Our All may be thus be regarded as something like a cross-section of a portion of the metacosmos, or a projection, as an x-ray picture gives a projection of part of the body. It is suggestive that our space is complicated and creative, with an elaborate geometry; pullulating with virtual (not directly observable) particles, it is much more than the mere absence of matter. It would seem natural that a hyperspace should be correspondingly more complicated and powerful, endowed with indefinite potentialities of order.

Human knowledge is approaching the stage where we should be able to learn systematically something of the metacosmos. Our knowledge of it may be inferred from its creature, our cosmos, and the intelligence that reflects it at the highest level. The mind can go on, through its order-building capacities, to explore directly something of the metacosmos, as in mathematics—something but certainly not all, as the full meaning of the metacosmos must remain forever beyond the powers of human minds. As the ultimate source of reason, the metacosmos should be partially accessible to reason; it is presumably also related to the mystical experience. The metacosmos also implies that there can be no total understanding of the cosmos.

The metacosmos is supernatural in the sense of standing over nature, but it is not unnatural. To the contrary, it is the essence of the natural and rational, like the laws that belong to it. Although made apparent through material things, it cannot be conceived as anything concrete; concreteness implies pertaining to material substance. It is akin to what scientists have in mind when they say "Nature is ingenious," or speak of "Nature" as something like an abstract God whose ways they admire and would fathom.

The metacosmos includes everything unbound by time and space, everything that, to our knowledge, is not accidental or fixed in particular material realities. This includes mathematical truths, logical relations, fundamental facts such as conservation laws, and the forces or potentialities that brought the universe and ourselves into existence and make possible the continual growth of order into the creative mind. Presumably the relationships and rules governing the material world known as laws of nature also belong to the metacosmos. Conceivably, however, the constants relating them concretely to phenomena are materialized, so to speak, as needed. We do not know how far the laws of nature peculiarly characterize this cosmos of ours, or how far they may be essentially mathematical, inherent in the metacosmos.

Perhaps many of the equations and basic facts must be as they are, although the seemingly arbitrary relations between physical constants may be restricted only by the requirement that they produce a universe capable of self-observation. It is mysterious whether there is some significance in such numbers as the ratio of masses of proton and electron, or whether they simply crystallized, so to speak, soon after the beginning.

Any cosmos coming out of our metacosmos supposedly must have much in common with ours, although we do not know to what extent laws of nature are general or contingent. It is fruitless to ask where the metacosmos came from, or whether there must be a metametacosmos, and so forth, permitting different mathematics and basic structures in daughter metacosmoses. It is interesting but useless to speculate about such a hierarchy of levels and generality in creative capacity.

Thinking in metacosmic terms suggests that all broad truths are related, although in ways perhaps beyond powers of understanding, from the basics of physics to the upper reaches of psychology and philosophy. It may be surmised that there is something in common in the generation of the physical order of things, the unfolding of life, and the blossoming of intelligence, culture, and civilization. The metacosmos is much more than a rulebook guiding the cosmos; it must contain drives or tendencies leading to the generation and flowering of our cosmos. It can be regarded as potentiality; material substance makes explicit something of what is implicit in the metacosmos, as the abstract is realized in the concrete. The deepest questions we can confront have to do with the relationship of the material universe with the metacosmos, in terms of which we have to seek partially to unravel the enigmas not only of physics and cosmology but of biology, psychology, and culture.

A creation is, of course, the thesis of all religions; it is rather sensibly assumed that there must have been something like a beginning, and one of the chief attributes of the Supreme Being is the fashioning of our world. However, the emergence of the cosmos from the fertile void does not imply that the creative force, whatever its nature, intervenes in individual events or takes any "interest" in human affairs.

Origin in something exterior to our material universe implies not a personal Creator as we conceive personality, but something grander, the qualities of which are manifested in its product, somewhat as an artist's self enters his composition. It is reasonable to suppose that the universe could come into existence only within a framework containing something like will and awareness or subsuming these qualities on a higher plane. Some physicists would see the eventual emergence of mind as necessary for the origin and existence of the universe;[9] it is less imaginative to contemplate that the universe had to be produced by something like mind. It is easy to believe that there should be analogs of intelligence and purpose in that which brought about a universe containing intelligence and purpose. As Paul Davies has it, "No one who has studied the forces of nature can doubt that the world about us is the manifestation of something very, very clever."[10]

In sum, something that had an inception exists in relation to its substratum. A constant eternal universe can be conceived as the whole of reality; a created universe is to be understood, so far as it may be understood, as shaped by a larger reality.

## Design

A related fact is that the universe is as though intended for us. Science has given a different and sounder turn to the age-old argument from design, the philosophical-theological claim that the wondrousness of our world must imply a Designer. The "anthropic principle" starts with the undeniable fact that we are the mirror in which the cosmos, so to speak, views itself, and that the universe must satisfy many exacting conditions (assuming the laws of nature) in order that thinking beings like ourselves could come into existence. There are doubtless other difficult conditions of which we have no knowledge; those of which we are aware are so demanding that, so far as probability can be applied to universes, ours is practically impossible as a random outcome.

We do not take for granted that the builders of a technological society have to be quite like ourselves. But intelligent beings are very different from nonliving matter, and the only reason that we have for believing such oddities to be possible is our own existence. It is hence quite logical to suppose that the more a material structure differs from ours, the less likely to furnish the requisites for intelligent entities and the self-observation of the universe. So far as we can reason, the existence of thinking beings implies a universe much like the one that we know.

It is to be assumed that intelligence presupposes creatures composed of many different parts in more or less stable but very complex flexible relationships. This means moderate temperatures and aggregate bodies—no thinking can emerge from a swirling gas. It also requires a flow of energy, something like a cool satellite in the vicinity of a hot body such as a star. Many other things have to be exactly right, within very narrow margins, to make possible large, complicated, stable, and yet alterable molecules, permitting the formation of intelligent beings. For example, the universe has to be of approximately the observed immensity in order to permit the rise of a civilization.[11] There had to be an excess of a mere one-billionth of matter over antimatter in the original fireball to leave behind the substance that would make up the stars and ourselves (after the near-total mutual annihilation of matter and antimatter).[12] If the rate of expansion of the universe had not been just so, within an infinitesimally small factor, galaxies could not have formed without collapsing.[13] A universe of black holes is inherently far more probable—by very many orders of magnitude—than one of bright stars.[14] Nuclear forces and the decay constant had to be precisely correct to permit the production in suitable proportions of hydrogen, helium, and heavier elements. For the stars to act as stable furnaces, relations of gravity and nuclear forces had to be perfectly attuned, despite their almost unimaginable disparity, with an exactitude virtually impossible to achieve by chance.[15] If parameters were slightly different, no supernovas could implode in seconds and eject the makings of planets. Many other basic relations had to be as they are, within extremely close tolerances, to permit the complex compounds and interactions leading to living matter.

The set of possible universes is, of course, entirely unguessable; but if every parameter were allowed to take any possible value—up to nearly infinite numbers—the probability of our Garden of Eden must be virtually zero. Finding a needle in a haystack is a certainty compared with the unlikelihood that a universe with random parameters would be capable of producing thinking beings.

We must hence suppose either that this universe was purposefully structured for the development of intelligence or that it is the lucky one of a near-infinity (or infinity) of chance-directed universes with all combinations of parameters. The former seems the more reasonable, or at least the more thinkable. The idea of an infinite number of universes being set in motion would be quite appalling if the mind could encompass it. It is extremely awkward to invent numberless universes to account for this one, which is quite spectacular enough.

The origin of our cosmos from something larger or higher does not, of course, prove the existence of any humanly understandable final purpose in it. One should not think in terms of a cosmos oriented toward the generation of ourselves. It is more sensible to hypothesize that a metacosmos with the capacity and propensity to create a material existence would give rise to an order-building cosmos capable of indefinite creative complexity and of self-awareness, which happens to be incorporated at this moment in our species. The uniqueness of this creation may, then, tell us much about the metacosmos.

We are not to assume that our not wholly admirable persons are necessarily the purpose, much less the final purpose of a universe that apparently will continue to exist for many billion years. But if our material being was made possible by something like design, the implications are large. If the cosmos was ordained for the generation of intelligence, we surely have an important role.

## COMPLEXITY

A relatively simple universe, we may assume, is more probable than a very complicated one; yet our cosmos is vastly, inconceivably complicated. If it is designed, it could well be called overdesigned, unless complexity is an inherent and essential part of its nature, necessary for the generation of self-awareness.

Simplification is the ideal of science, which seeks to tie things together, finding common causes and explaining complicated things in terms of more basic ones. But in practice, to "simplify" does not mean to make more easily understandable, only to bring together conceptually; and complications arise and multiply like hydra heads. Early in this century, it was believed that the world consisted of matter plus fields, and that matter was made of electrons, protons, and neutrons. The goal of unification seemed at hand. But ever more "elementary" particles, up to the hundreds, were found or made. It turned out that the heavier particles were nearly all composed of different but inseparable parts, whimsically named "quarks." This was again promising of simplification, and seemed to represent the end of a road, because quarks cannot be knocked out of a proton and studied in isolation. But it was found that there are six quarks, each coming in three "colors" and each with its antimatter counterpart, plus six leptons, or members of the electron family, each with its respective antimatter particle. In addition, a large number of particles mediate interactions or forces, beginning with eight kinds of "gluons" holding quarks together to form protons, neutrons, and their relatives. Why all these particles exist and have the ratios and interactions they display is mysterious. As if known particles were not enough, efforts to generalize physical laws and particularly to account for the beginnings of the cosmos postulate varying numbers of as yet undetected and perhaps undetectable particles. One theory of "supergravity," intended to simplify matters, requires 70 different basic particles, a large majority of which are only gleams in the eyes of the equation-makers.[16]

Non-physicists can experience the complexity by seeking to penetrate it, by trying to digest even much simplified explanations of the experts. For the layman, rather elementary equations might as well be written in Egyptian hieroglyphics. Physicists like to call their beloved equations "beautiful," and a few spare symbols are said to enfold the greatest mysteries. But years of study are needed to understand the meanings of the squiggles.

The relationships of the numerous entities are all guided by quantum mechanics, which is a most beautiful and successful mathematical-theoretical formulation of the interactions of particles. Its basic ideas are fairly simple, but there are many complications, some still controversial. In application, it is forbiddingly involved; many thick books are needed to set forth what is known of the behavior of the supposedly simplest things.

An electron, for example, is regarded as a dimensionless point with no structure. Yet it, like all components of the universe, has a wavelike as well as particle-like nature, travelling as a wave but interacting as a concrete point-entity. It continually generates and interacts with short-lived virtual photons and other half-existent and unobservable particles for an indefinite distance around itself.[17] It may be conceived as a quantum of the electron field, just as a photon is a quantum of the electromagnetic field. It has a fixed charge, mass, and magnetic moment, the magnitudes of which are not known to be related to anything else. Like all particles, it has a fundamental property called spin, like a twirling top, with a certain direction and orientation, although electron spin has no counterpart in the macroscopic world, and nothing is postulated physically to be turning on an axis. Unlike a gyroscope, the spin cannot have a varying direction but must be either positive or negative, and its axis must turn over twice in order to return to its original position. Particles such as the electron and nucleons have half-integral spin (values of 1.5, 2.5, etc.), while interaction particles, such as the photon, have integral spin numbers, a difference that is basic to the makeup of matter.

The electron has complex relations with other particles and with photons and is capable of intricate maneuvers in the field of an atomic nucleus; the difference between oxygen and sodium arises mostly from the different configurations of their electrons. Electrons play many tricks, such as interacting with other electrons at great distances (in electron dimensions) despite intervening material, and flowing frictionlessly in superconductive substances. It is a token of the perversity of nature that, long after science had seemed to have reached a fair (although incomplete) understanding of superconductivity near absolute zero, higher temperature superconductivity proved utterly baffling, an empirical discovery without a theoretical basis. In various ways, electrons and other particles defy common sense. Although, like photons, they register at a point, they spread over large areas and can follow more than one path at once. The quarks that make up nucleons seem even more enigmatic. The mathematics of gluons is too difficult to permit useful calculations relating theory to experiment.[18] It is difficult to imagine how all this information can be contained, so to speak, in dimensionless near-nothings. The obedience of particles to the rules and principles of quantum mechanics is somehow inherent in the order-potentialities of the cosmos.

The universe is far from describable in terms derived from ordinary experience. Although regular behavior emerges statistically from larger numbers or masses, the behavior of atoms and smaller things is never fully predictable;[19] and small uncertainties grow rapidly, in dynamic interactions, to large differences of outcome. This indeterminism is not a matter of inability to measure but is fundamental to the nature of matter; the universe might well be said to consist of quantum fields[20] governing probabilities. An atom does not really have orbiting electrons but orbital wave functions telling where electrons may be. Some physicists long ago concluded that the problem of measurement implied consciousness or mind,[21] and they went on to postulate that things are real only so far as observed—without being able satisfactorily to define "observation." This leaves the theological explanation that everything becomes real as observed by God—while some theologians believe God is made real by human apprehension.

Even good explanations carry deep difficulties. Standard quantum theory is enormously successful and true in a very real sense. It accounts for a great variety of facts, is refuted by none, and has made possible many a modern marvel, from lasers to computers. But it suffers from underlying mathematical arbitrariness,[22] and it is disconcertingly difficult to harmonize with General Relativity, which has also been highly successful and seems fundamental to understanding of the universe.[23] Scientists struggle with such puzzles as the apparent fact, shown by the motions of galaxies, that 9/10 or more of the matter in the universe is invisible.[24] They have recourse to fantastic theories; for example, the ultimate components of particles are conceived as loops in ten or more dimensions (all but three of the spatial dimensions being so tightly coiled as to be unobservable), as much smaller than a proton as a proton is smaller than a racetrack. It is also theorized that there may be "superstrings" stretching across the universe, outrageously energetic and massive threads quite unlike all known matter, the only evidence for which is that they might solve sundry riddles of cosmology.

Theoretical physicists find it hard to follow the mathematics of recent theories intended to simplify physics by bringing all forces into a common framework.[25] Theoreticians want to generalize and harmonize ever more phenomena, but they usually do so at the price of adding to the complications of the explanation, with more conditions, parameters, hypothetical entities, and mathematical procedures, the necessity of which needs explanation as much as the phenomena that the equations should elucidate. If the electromagnetic and weak forces are brought together conceptually, this is satisfying but hardly a simplification, as new specifics must be assumed to account for the differences. Simplicity is like a rainbow, retreating as one approaches. It is beloved of physicists precisely because their work is so outrageously complicated.

Space itself, as analyzed by Einstein, is almost unimaginably involved. He was enamored of logical neatness and beauty, and his theory of General Relativity was intended to unify inertia and gravitation. The fundamental premise is simplicity itself: that all the laws of nature, including the speed of light, are invariant regardless of the movement of the observers. However, as often happens, the simple principle leads to appalling complications. One outcome is that time flows at different rates for different observers, depending not only on their motion but on their gravitational field. The theory is beyond the understanding of any but the best prepared. The 10 tensor equations needed to describe space are difficult even for physicists; worse, they are solvable only for a few of the very simplest cases. Einstein himself made a serious error in his initial calculations of the bending of starlight by the sun, and the general equations for the universe were solved only long after.[26] Whatever the difficulty of General Relativity, its explanation of gravity possibly represents an oversimplification, as evidence has been found of short-range deviations from expected gravitational force.[27]

Not only is the relativistic geometry of space forbidding; it appears that apparently empty space is a continually bubbling stew of ghostly "virtual" particles, which borrow sufficient energy from the vacuum for a brief existence and exercise potential effects on "real" particles before quickly vanishing. Moreover, quantum events seem to occur instantaneously over large distances; the very nature of space becomes questionable. According to accepted theory, if there exists a single magnetic monopole it must have effects across the universe. A minor mystery is that particles distinguish right from left. Contrary to the commonsense assumption, particle interactions are not entirely symmetrical to reflection in a mirror.[28]

Physicists always make unexpected discoveries as more powerful accelerators raise particles to higher energies. The quest seems to have no end. There probably can be no complete physical theory, no "formula of the universe," just as there can be no complete mathematics, if only because the material universe is not self-sufficient: its rules are not all contained within it.

The cosmos is vastly more complex than necessary to produce the physical universe as we know it. For example, only two of the six leptons and two of the six quarks are needed to make the physical universe,[29] and most mediating particles would seem to be superfluous. If ours is merely a (or the) universe capable of evolving intelligence, a simpler one with fewer parameters would seem to be far more probable. However, complication may be the most general law of nature. In the realm of the smallest and simplest components of matter, there may be an indefinite number of independent givens, from the masses of elementary particles to the equations of Relativity and the details of the profoundest theories of the near-unknowable.

Complexity arises equally at the level of the very big. Astronomers prior to the discovery of quasars in 1963 and of pulsars in 1967 thought that they knew all the species of bodies that could exist in the heavens; recent observations have turned up mysteries from the lumpiness of clusters of galaxies to unexplained sources of gamma radiation. The manifold phenomena of turbulence and "ordered chaos" that occurs in all manner of systems, from ocean waves to the winds, defy ordering. Much less can one expect to find simplicity in the inherently more complicated interactions of compounds. Even at the molecular level, we find many almost independent realms of knowledge; the chemistry of polymers, for example, has only fundamentals in common with the crystallography of metals. Living nature is, of course, far more complex. As one proceeds to higher levels of order, to psychology and culture, with the overwhelming complicatedness of personality, mind, and society, the hope of finding simple explanations to tie all facts together becomes less and less realistic and more and more an impediment to understanding. With the aid of a computer, one can go on to evoke new intricacy ad lib, as in the beautiful and infinitely varied shapes that computers make from extremely simple iterated equations (Mandelbrot and Julian sets and their relatives). Fantastically elaborate forms are as though conjured from nowhere.

Indeed, hardly anything is simple except so far as the scope of attention is limited; one claims to understand the hardboiling of an egg by ignoring the processes inside the shell. The higher the level of being, the more understanding must remain partial and qualified. Statements about human society can seldom claim incontrovertible truth; there are no real "laws" of political science. A rising degree of complexity seems inherent in the nature of the cosmos, which we may see as reflecting the metacosmos: there are always new facts and interrelationships. This open-endedness is repugnant if one dreams of bringing everything together in a simple web; but complexity is the brother of creativity, and the end of discovery would be the end of development in the universe, as it would be a sort of mental death.

## DUALITY

Because of its inception from a generative substratum, the cosmos has dual aspects: the material particles that came out of the fireball and the patterns of ever-growing interactions. There are on the one hand material substance (endowed with energy) and on the other the elements of order—not only the regularities of physical interactions that make stars and snowflakes but the shapes of life and the logical framework of things, down to the convolutions of mathematics and the mind in which mathematics takes shape. The two aspects overlap and interact to make the configurations of reality.

The matter that makes up the sun and earth, the desk and the lamp made luminous by electricity, and the neurons that direct the hand to write are like the paper and ink of which a book is made. But the essence of the book is not so much paper as words, and reality is not only substance but form. A house is built of bricks, lumber, plumbing, and so forth; but equally important is the design governing how the house is built, the way in which the materials are put together and the purpose they serve. The materials are only the means of fulfilling the plans.

On the highest level, we see this duality in the antithesis between mind and matter, between observer and observed, between thought and the material brain. A human being may be perceived either as a thinking, feeling entity or as a highly organized aggregation of material substances. One may communicate with a person and meet a mind, or dissect a brain and find a tangled mass of fibers.

Yet there is no sharp dichotomy. Form or structure exists only as incorporated in material substance; the mind depends upon the physical brain. At the same time, matter seems almost automatically to take on elements of order whenever it can, as a molten metal forms crystals when it solidifies. Order and substance are opposite but inseparable. Elements of order do not leap out of nowhere but grow on their material foundations. The duality of the universe is not that of two separate principles but of opposite vectors, like the two ends of an arrow. This principle is very broad. The dichotomy of classical logic, "A or not-A" is convenient for simplicity and rigor of thought. But it is usually truer to say "A or not-A, or part-A, or qualifiedly-A, or perhaps-A."

Matter itself is unsolid. Only protons and neutrons have any considerable mass, the electrons weighing only about 1/1800 as much. Photons and (presumably) the ghostly neutrinos have no ordinary mass at all, only its energy equivalent. A large majority of particles, mesons and baryons (heavier particles) other than the protons and neutrons, are utterly evanescent. Particles themselves have a dual nature, wave-like and particle-like; that is, they carry information and substance. Electrons can be annihilated by positrons, leaving behind only a burst of energy; like other particles, they can be created from pure energy. The virtual particles that play a key role in the properties of matter can hardly be called material in the usual sense of the word; they can be neither trapped nor directly detected. They exist, in effect, to enable physicists to describe the interaction of particles. In a sense, all particles are not so much entities as manifestations of the rules and characteristics that are integral to their nature. The universe may be said to be composed less of material things than of processes and events.[30]

Which aspect one prefers to emphasize depends on taste, philosophy, and the kind of questions at hand. In everyday life, there are materialists and idealists, hedonists and ascetics. Similarly, there are materialists and idealists in science and metaphysics, or reductionists, who seek explanations by taking complexity apart to examine components, and holists, who try to find broader meanings in complex assemblies. In sociology, some seek understanding primarily in terms of material conditions and needs; others, by examining traditions, myths, and customs. Ironically, many of those who study that most ethereal of entities, the human mind, insist on regarding thoughts and feelings as the outcome of physical (neural) structures and events. But many physicists dedicated to plumbing most deeply the nature of matter stress the abstraction of their results and the crucial role of the observer.

As the universe expands not only in space but in information, the aspect of order gains importance relative to the aspect of substance. If the universe began with undifferentiated substance, the interactions of structures have increasingly determined events. In a degree this is true of astronomical bodies, from the variety of galaxies to the paradoxes of black holes. It is more clearly the case with living creatures, in which an extreme degree of organization of matter produces incredible adaptations. It is most strongly evident in the operations of intelligence and the mind, which are like apparitions shimmering on the surface of the material substratum but have the power to create or destroy on a godlike scale. The highest order of material things thus partakes of or approaches the nature of the metacosmos.

## CREATIVITY

The birth of cosmos from metacosmos was a gigantic leap into complexity; differentiation and order began in the expansion and cooling of the fireball, and they have been proliferating ever since. At the earliest time concerning which physics can fruitfully speculate, there seems to have been only a single kind of particle, governed by a single force; the different particles we know and the four present forces (nuclear, electromagnetic, weak, and gravitational) are believed to have been equivalent at an unimaginably high energy level. There was no structure of any kind, only a featureless chaos.[31] The universe contained almost no information except its temperature and density.

But material substance is as though endowed with a great capacity for self-organization.[32] Very soon there began what the physicists call "loss of symmetry," which means that, for reasons unclear, things took on different characteristics and more complex symmetries. The forces diverged and came to differ tremendously in strength, range, and modes of action. The particles separated into electrons and quarks, the latter joining, as soon as they had cooled slightly, to form protons and neutrons. After nearly all matter was destroyed by antimatter, the remaining one part in a billion came out about 75% hydrogen, 24% helium, and 1% deuterium and other elements. After about 100,000 years radiation was decoupled from matter, and the universe was sufficiently expanded and cooled to form stable atoms. The gas became ever clumpier; in a hundred million years, motion was causing turbulence, and clouds were beginning to form protogalaxies. In about a billion years, galaxies were condensing into stars —stars are social, always belonging to galaxies or clusters.[33] When their thermonuclear processes had proceeded sufficiently, some stars exploded as supernovas, spewing heavier elements into space and making it possible for solid planets to form around stars like our sun. On the cool earth, atoms fitted themselves into large molecules, which made possible living creatures. Some of these eventually attained sufficient complexity to build a new type of order, human culture and civilization.

There has thus been a continual increase of order embodied in material substance. This defiance of the Second Law of Thermodynamics (according to which there is an inevitable and irreversible increase of entropy or disorder in any closed system) is made possible by the openness of the system, that is, its expansion. In cooling and growing, the cosmos has tremendously increased its content of form, structure, and information. Each level of order builds on what has come before, a process that has brought about ever more effective means of generating new order out of turbulence and disorder, culminating in the inventive autonomy of mind.

It is absurd to suppose that all the potentialities of the universe were somehow written into the featureless substance of the initial fireball. The bricks are not born bearing the design of the house. New order continually arises from unpredictability, as the turbulence and disorder of material things interact within regularities that guide them. Random inputs selected for stability, or chaotic and unpredictable turmoil guided by the laws of nature, have led to the formation of galaxies of their various sizes and shapes, the making of complex compounds, the evolution of life, and ourselves and our society, in fulfillment of the potentialities of the metacosmos.

The cosmos is to be seen not as dead and mechanical but as something alive and developing, like a giant tree coming out of a seed, or a tiger growing from an ovum; something very complex emerges from a small and seemingly simple beginning. The seed, however, contains in its molecular structure the instructions that direct the formation of roots, stem, leaves, and flowers; and the nucleic acid of the ovum carries the blueprint permitting the development of teeth, muscles, guts, nerves, and fierceness. The cosmos, to the contrary, began in complete simplicity and acquired patterns as it grew.

In the sequence from quarks and electrons to chemical reactions, living processes, and the human mind and social behavior, whatever appears at any stage is based on what went before but builds on it in ways exceeding any possibility of calculation. When things are combined, there are results unpredictable from study of the parts.[34] A pile of bricks tells little about what can be done with arches, and the turbulence of the surf cannot, even in theory, be predicted in detail from knowledge of masses and forces.

At all levels, the whole is more than the sum of parts. As one proceeds to higher levels of order, with new concepts coming to the fore, there is an increase of something like autonomy. The electron is governed by its equation, but the outcome of molecular combinations is more broadly unpredictable. The inception of weather patterns is not even theoretically predictable, because a tiny input, such as a trivial air current—the flapping of a butterfly's wing is the conventional example—may upset an unstable condition with indefinitely expanding effects, atmospheric movements culminating in a huge storm.

Levels of order are built one atop another, compounding complexity and unpredictability. At each level, the laws of lower levels apply fully, but new integrations come into play, and higher-level laws or regularities appear. Thus in living matter, we assume that atoms and electrons obey all the rules; but the system takes on behavior unforeseeable by the laws governing its components. This is equally true of molecules and people. With rising complexity, knowledge becomes less exact and verifiable, and prediction becomes more difficult, from physics to biology to psychology to sociology.

Such a simple interaction as the forces governing more than two bodies under Newtonian gravity cannot be described with mathematical exactitude, and the stability of several orbiting masses may be impossible to calculate. Even though a new reality may be theoretically understandable from simpler things, the relationship is likely to be excessively complicated. For example, it is said that some 300 quadrillion calculations are necessary to calculate the mass of the proton from knowledge of quarks and their interactions (chromodynamics).[35] It is possible to derive from knowledge of neutrons and protons only a rough idea of the binding energies of nuclei (which determine their reactions in stellar interiors) or the half-life of radioactive elements. What can be told of electrons, protons, and neutrons gives only a general idea of the characteristics of atoms. Science is still learning about the behavior of the simplest of substances, hydrogen, which assumes unforeseen properties in solid form. Knowledge of hydrogen and oxygen gives only a beginning of an understanding of the complexities of water, from its tensile strength to the temperature to which the liquid can be chilled and the several phases of ice. The qualities of an alloy are not exactly knowable from those of the metals composing it.

It is not possible to write down the quantum equations of organic molecules, much less solve them. What may come out of the combination of complex substances appears in the infinite creativity of life, leading to the new universe built by intelligence. The increase of order becomes ever more subtle and abstract, and there is no sign that this progression has ever to come to an end.

Even apart from quantum uncertainty, determinism is quite untenable, and effects are greater than their causes. Information is perennially coming into existence; and it is also being destroyed, as when a lump of sugar is dissolved or a page is burned or a creature dies. Could the future of the universe be predictable by a superintelligence possessing total information of its present state? This would be contrary to the nature of the cosmos. One would in any event prefer to assume not, because a predictable universe with a foreseeable outcome would be idle and uninteresting. If past and future were preset, like the drama frozen in a movie reel, the universe would be meaningless. Effect follows cause closely enough to make a consistent world, but the human desire for order is the only reason to suppose that the future is written in the present. It is thus with the weather, biological evolution, and human affairs: stability, continuity, and near certainty or high probability of major near-term outcomes are joined with change and unpredictability to make details and distant outcomes unknowable. Events and forms are restricted by preceding events and conditions but not fully fixed by them; snowflakes have to be hexagonal but vary infinitely and partly unpredictably in detail, while no one can tell just what blizzards are going to be until they are done.

Einstein, uncomfortable with unpredictability, used to say that God did not play dice with the universe. But the freedom of incomplete causation is not frivolous. It underlies the creativity that brings new order into being; it is a higher act to create something with creativity of its own. Greater complication presupposes a higher order, and the higher order is simply that which has the greater capacity for new creativity. The basic unpredictability of quantum mechanics presents a microcosm of the universe; within constraints, the details spell themselves out. The indeterminism of outcomes is a component of the cosmos, like matter and order. Above all, the dynamic mind makes its own future: we are partly what we have to be, partly what we want to be.

The generation of new order by this creative freedom may be the deepest characteristic of the universe. According to a Talmudic text, God said, after making the world, "Let's hope it works."[36]

## PURPOSE

Probably only a universe with an inherent tendency to the creation of order and complexity can become self-observant, as ours becomes through us. Conceivably the ability to build new order is necessary for the existence of a cosmos within our metacosmos, without which it would collapse or die of inanition. At least, the cosmos that is to bring about a partial reproduction of the metacosmos must have a powerful tendency toward increase of complexity; that is, it must have a clear-cut direction, correlated with its expansion.

Although the universe cannot be said to have a purpose in human terms, it is going somewhere; and a teleology has been valid in the large, though not always in the small.[37] There may be reverses, as there have been reverses in the progress of civilization—dark ages after eras of innovative prosperity—but local reverses give way to further and broader advances.

The mixture of order and disorder in the cosmos makes possible a history and a drama. It is its grandeur to generate ever new, more complex and interactive things, to bring about order capable of responding to the world in a systematic way and producing still higher integrations. This surely has to do with whatever purpose it may have. So far as we know, we are the highest expression, the present climax of that ongoing process, although certainly not its finality. Whether or not we merit congratulation as the expression of cosmic necessity, we correspond to the general expansion of non-random arrangements, especially systems productive of further order, the acme of which is self-aware intelligence.

The physical insignificance of humanity is irrelevant. We are properly awestruck by the dimensions of our surroundings; but if the universe was generated by something like an intelligence, its most understandable purpose would be something like the reproduction of some of its essence through the generation of new order and intelligence. The purpose of the cosmos may be to do what it seems to be doing.

2

# The enigma of evolution

## Emergence

If the materialization of the cosmos was the first miracle, the second was the generation of life and the incorporation of purpose in organisms. Inanimate nature produces surpassing beauty in fantastic diversity, from rainbows to galaxies; but the nonliving has no direction of development. A crystal grows merely by molecules falling into ranks; its order is static. A living organism, to the contrary, is structured for function; it is endlessly variable in detail and capable of change and development. To be living, something must be set off from its surroundings, becoming a new and partially independent entity, although (except in an unlifelike state of dormancy) it is constantly exchanging materials with the environment. It must also have capacities of adaptation and response. But what most of all distinguishes the living from the nonliving is the ability to generate more life, with the closely related capacity for self-repair. Living things make their future and project their likenesses; in them, purpose becomes the essence.

Life is problem-solving.[1] Why does a planet or a grain of sand have a certain shape? This can be answered only in terms of the history of the planet or sandgrain, not its uses. But a paramecium has cilia and mitochondria for locomotion and metabolism, just as a leaf turns toward the sun in order to catch more of its rays. Flowers blossom to attract pollinators, and a hawk's superb eyes serve to spot mice from far above.

The cosmos did not hurry to complicate itself with organic purpose. Billions of years of preparation were needed to set the stage, and life could not be born until after stars had cooked heavier elements in their interiors and erupted as supernovas. Dust clouds formed in this way provided the substance of new stars, such as our sun, whose age is only about one-third that of the universe, and of planets to dance around them.

Any lifelike development will surely be based on carbon because it is the element best able to link up in long chains. Living material consists of carbon chains with attached groups of atoms of hydrogen, oxygen, and nitrogen. These four elements, which are the commonest of the universe (except helium and neon, which do not form compounds), comprise over 98% of the human body. Life also uses phosphorus, sulfur, sodium, iron, calcium, and traces of more than a dozen others, all of which are present in the ocean in which life began.

Yet if a disembodied intelligence contemplated a lifeless earth, it would require much imagination to guess that the muck of the tidal pools could ever organize itself into the maker of skyscrapers and atomic bombs. Most biologists find the spontaneous generation of life unlikely, however favorable the physical and chemical conditions. However, precursor-compounds of living substance, including amino acids, are formed with surprising ease. In 1953, a graduate student found that passing a spark through a mixture of methane, ammonia, and water vapor (the substances believed to have composed the bulk of the earth's early atmosphere) produced in a few hours a rich harvest of building blocks of protoplasm.[2] Large quantities of similar compounds, including all five nucleic acid bases and nearly all amino acids, have been detected in the dust clouds of interstellar space or in meteorites. It has consequently been speculated that the beginnings of living things were rained upon the earth.[3] There is no real reason to believe this, however; and the hypothesis that a spore somehow drifted from outer space only shifts the problem of the origin of life to an unknown elsewhere.

It is more conventionally guessed that the inception of life may have been associated with clay crystals, to which organic molecules adhere.[4] They also have catalytic properties, which might be energized by volcanic thermal vents or ultraviolet radiation in tidal pools. The development of life must have been an extension of the ability of crystals to grow and in a sense reproduce in a suitable solution. Innumerable compounds of carbon, oxygen, hydrogen, and nitrogen would be precipitated from the early ocean. As soon as certain molecules or small clumps of molecules were able to promote the precipitation of more large molecules or clumps like themselves, something approaching life would have been on the way. The distinction between living and nonliving is not totally clear-cut, and there no doubt were enormously complicated molecules and interlocking chains of molecules long before there was anything that could be called living.

Very soon after life effectively began, it must have increased exponentially. Expanding rapidly across the virgin seas, it changed the environment by consuming stored-up nutrients of the ocean broth, thus precluding new life-beginnings. It must be for this reason that life on earth is a single web, probably with a single molecular origin, like the origin of the universe at a point. This is indicated by the similarities of basic chemistry of all creatures. They use the same 20 amino acids, reproduce by very similar nucleic acid chains with the same arbitrary code, have the same chief energy carrier (adenosine triphosphate), porphyrins (used in respiration and other processes), and so forth.[5] It may also be significant that all natural amino acids are optically left-rotating, although right-rotating isomers can be made with equal ease. The fact that all living things have retained such biochemical similarity suggests either that there may not be many ways of carrying out the fundamental processes or that any very different ones are difficult to achieve or are noncompetitive.

There undoubtedly were innumerable stages on the way to the fully living. Bits of jelly hardly separable from the pristine mud must have merged into tiny globules, which organic materials form quite easily. Some of these would become capable of absorbing materials dissolved in the waters to make more little globs like themselves, eventually giving rise to bacteria-like ancestors of plants and animals.

No one knows which of the two basic components, nucleic acid or protein, came first. Perhaps they somehow came into existence jointly, or something simpler may have produced protolife that was able to initiate the generation of substances so complex as proteins and nucleic acid. Short strands of the more primitive species of nucleic acid, RNA, can act somewhat like proteins and also replicate themselves; a few of these, joining with some enzyme molecules, may have formed predecessors of cells. When they managed to develop an enveloping membrane the battle was nearly won.[6] The fundamentals must have been invented well before the earliest fossil traces of bacteria, which date to about 3.5 billion years ago,[7] a few hundred million years after the earth became habitable. Curiously, the tremendous leap from the inanimate to something much like modern bacteria and blue-green algae apparently required only about 300 million years, but there was little further change for 2 billion years afterward.

At first, life was limited to chemical energy sources, like the sulfur compounds on which some modern bacteria feed where hot springs bring them up from deep in the earth. With the appearance of photosynthesis to use solar energy, the biosphere must have increased hundreds of times over. Life first changed its environment by consuming the materials that gave it birth; the second great alteration was the gradual accumulation of a highly reactive element, oxygen, in the atmosphere, and the concomitant consumption of insulating carbon dioxide. The atmosphere became something like an extension of the biosphere.[8] This made possible the extraordinary stability of earth's climate, as the replacement of carbon dioxide by oxygen compensated for the increase of solar radiation, keeping temperatures moderate for about four billion years.

Lack of oxygen was probably essential for the inception of life, because in its absence many complex oxidizable compounds could stew in the oceans. But the gradual accumulation in the atmosphere of oxygen from photosynthesis made possible much more active life with more energetic chemical processes—large quantities of oxygen are continually being consumed, and the present supply of oxygen would be exhausted in a few thousand years if not continually replenished by plants.[9] Larger and better organized single-celled organisms with nuclei (eucaryotes) arose after some 2 billion years. They apparently came about through symbiosis; what had been independent bacteria-like organisms became organelles of larger and more complex cells. Such organelles, mitochondria and the chloroplasts of plants, reproduce as though they were bacteria embedded in the protoplasm. This cooperation and the segregation of genetic material in a nucleus gave a whole range of new capabilities.[10]

The more elaborate organization of nucleated cells incidentally made possible a regularized mixing of genetic material, that is, sexual reproduction. It also permitted specialization. At a lower stage, as in slime-molds, organisms that are unicellular for most of the life cycle come together and cooperate with special roles in reproduction. At a higher stage, cells become permanently specialized according to their position in a large permanent aggregate of cells. Multicellular plants and animals appeared about 700 million years ago; worms were followed rapidly by jellyfish, arthropods, and other phyla.

Animals developed hard parts about 600 million years ago, when there was a great and unexplained upsurge of new forms.[11] Then, in the Cambrian period, all the major divisions of living forms appeared in the book of rocks, the record of which is thereafter fairly complete. Animals followed plants onto the land about 400 million ago, thanks to the protection from ultraviolet radiation provided by ozone in the oxygen-rich atmosphere. The dinosaurs ruled for 125 million years, while small mammals lurked in the shadows. About 65 million years ago, the dinosaurs vanished for reasons unclear but perhaps associated with a celestial event such as an asteroid impact darkening the skies and causing a chill fatal to large animals. The mammals then came into their own, and *Homo* rose to unprecedented ascendancy.

## Directions of development

Nearly all evolutionary roads are dead ends, and almost all species that have existed on earth have become extinct; the story of life is mostly wastage. Biologists commonly see no general law of progress, no clear direction of advancement.[12] Yet we can find some broad directions in evolution above much randomness. There has been an increase of structural complexity in many lineages, as from early jawless fish to higher mammals. It is true that many species, such as crocodiles and horseshoe crabs, have changed little over eons. Some species, especially parasites, have lost structures and become outwardly simpler, although they have their own often very complex specializations. But they are exceptions. The usual rule has been to add to structures. Indeed, growth of complexity has apparently been greater than survival value would dictate; animals are in many ways more complex than they seemingly need to be to perform their tasks, somewhat as the universe appears to be far more complicated than it has to be in order to support such creatures as ourselves. One easily perceives this from any detailed consideration of the physiology of the best studied animal, *H. sapiens*.

Increasing complexity comes about partly because adaptations are not so much replacements of systems as something overlaid atop them, in whatever way happened to be accessible. Adding a new floor to a building is easier without tearing up the foundation. New structures are formed not in the most logical fashion but from the parts available in the given sequences. But no simple generalities explain why evolution should have provided the human brain with more than 50 neurotransmitters.

Complexity permits specialization, a common rule but by no means a universal one. Development of new structures and forms has meant an immense increase of variety, the bewildering, exuberant multiplicity of perhaps 30 million species of animals (including possibly 10 million species of insects and a similar number of mites), as though an artistic designer exulted in the creation of living forms. By continual experimentation, life has increased its capacity to adapt to different habitats and opportunities, or to occupy ever more niches, in the biologists' term, from crevices in Antarctic rocks to blistered sand flats and near-boiling hot springs, from the upper stories of rain forests to oceanic abysses. The total amount of life has grown immensely, ever more to cover the earth in an unending push into new environments. The base of the great pyramid of life has moved little, but the whole mass has expanded and the apex has risen.

Increasingly complex structures and forms offer ever greater possibilities of innovation. When creatures were simple and the elements of their organization few, evolution had little to work with and was glacially slow. With more structure, there are more possibilities of variation; as more elements are interwoven, there are more opportunities for new integrations from their recombinations and interactions. Life was only bacteria-like for some 2 billion years; but the course from the first amphibians to man took only 300 million years. Once the great reptiles had left the stage, mammals and birds rose and diversified like the amphibians and reptiles long before them coming onto the nearly vacant land. Only 30 million years lie between a tree-shrew and the reader. The evolution of man, which took some 5 million years from an ape with prehuman characteristics, was especially rapid in the last million years.

Yet the rule is not steady change but spurts, stasis followed by an outburst, perhaps released by an era of extinction clearing away a large majority of living forms.[13] There was an evolutionary explosion, unexampled so far as is known before or since, around 570 million years ago. Not only did all present major groups of animals (phyla) suddenly appear, but also a large number of weird-shaped creatures (to our perception) flourished briefly,[14] as though nature were trying out all possible variant plans for multicellular animals. There was a lesser radiation of forms when life conquered the land, and again when the dinosaurs vacated the stage.

While variety has increased, basic inventiveness has not; evolution, having fashioned all the phyla, or groups with a distinct body plan, has generated no new basic pattern for several hundred million years. Mammals and birds branched out exuberantly for tens of millions of years after the dinosaurs, but no important new groups have arisen in the past 30 or 40 million years. Insects in particular have remained virtually static for that time.[15] This is understandable because innovations are overlaid on basic structures, a new body plan would require long perfection to become competitive, and existing forms have occupied the accessible spaces. Now *Homo* has opened up a new field of culture, in which great innovations require no structural changes.

The reward of more complex structure has been ever greater autonomy. To be alive means to be set apart from inanimate matter and to possess new capabilities for change, over the eons to become more independent of the surroundings. Protolife became definitely living by the formation of a membrane separating protoplasm from the watery medium. It was a further step away from the inanimate when protoplasm became motile. Bacteria are mostly sessile, absorbing nutriment from the surroundings. An amoeba travels a bit, searching for food. A medusa idles with the waves, tentacles alert to nab its quarry. Fish chase theirs. Amphibians freed themselves in adulthood from restriction to water, and some of them have become able to live in deserts. Reptiles by enclosing their eggs made themselves independent of ponds and streams for reproduction. Fur or feathers and constant body temperatures enabled mammals and birds to function better in changeable climates. The erect mammalian posture means better locomotion; a deer runs much faster than a large lizard. Birds soared to the highway of the air. Hominids made themselves less subject to the environment with clothing and fire, perhaps before they could properly be called humans.

It is significant, on the other hand, that animals, while emancipating themselves from many external limitations, have become more dependent on others. Animals from fish through reptiles ordinarily deposit their eggs and leave the hatchlings to their own devices, but birds incubate and feed their young, and mammals gestate and suckle theirs. Baby monkeys need not only nourishment but the warmth of maternal care and the stimulation of play, without which they grow up psychological cripples. Dependency of human offspring is the most complete and the lengthiest of all, while young and old alike are ever more in need of the society and support of their fellows to function as humans or even to remain alive.

Living creatures have achieved freedom also by learning better to respond to things around them, acting purposefully in a more variable and controlled fashion. Primitive reactions are rather mechanical, vegetative, or chemical. The bacterium moves toward nutrients, the root grows down into the earth, or the vine coils about a post. The tissue receiving the stimulus responds to it automatically. At a higher level, signals are carried from one part or organ to another. The sensitive mimosa, touched on a leaflet, shuts a frond and bends away. Some simple animals, like the jellyfish phylum (cnidarians), have a network of nerves to carry messages around the body, causing contractions. Primitive worms have a little capacity to learn; and in higher animals, centralized controls make possible appropriate responses based on information.

The primary sense organ is touch, but animals have developed an impressive array of receptors to use many sources of data, such as electromagnetic waves in the range most strongly emitted by the sun, vibrations and dissolved substances in the surrounding air or water, and less commonly electric and magnetic fields. There evolved a central switchboard to receive the messages of the senses, coordinate them with experience or use memory traces (that is, learn), and send out instructions.

The integrating organ, the brain, intervenes between perception and response, first in simple and later in very complicated ways. It becomes the director of the whole body, although more primitive controls and the chemical message system of hormones (many of them produced by the brain) remain indispensable. The enlargement of the braincase and presumably of intelligence has been fairly continuous in various families of the vertebrate order; but it has been so slow (except in the human development), as to suggest that in ordinary circumstances it has not been extremely useful.

Elaborate instincts seem to be much more readily evolved than intelligence. Some creatures make phenomenally good use of a few neurons; for example, the honey bee navigates, keeps house, communicates the major facts of its existence, carries on a complex social existence, and has a language of sorts, all with less than 1/100,000 as many neurons as a human.[16] The more generalized capacity to use experience as guide to action requires a much larger number of cells, interconnections, and switches, as in mammals and birds.

In most groups the development of the biological computer has come to a modest level and halted. There is no indication that crabs, sharks, or crocodiles are any brighter than their ancestors of a hundred million years or more ago. But there has been a long-term tendency toward improved coordination of reactions, partly built into the organism as instincts (such as the patterned behavior of insects, in which learning is narrowly conditioned), and partly as intelligence (as in large mammals, such as carnivores, elephants, and higher primates), which increasingly takes command over innate directions and permits flexibility of response.

A major advance in responsiveness came with the invasion of the land. The more varied terrestrial environment, with greater possibilities of separation of populations both by adaptation and geographic isolation, greatly increased the variety of life, speeded up evolution, and raised the needs and rewards for intelligence. The reptilian brain enlarged very slowly, although some of the later dinosaurs seem to have had a rather developed social life. From reptiles to mammals, the average ratio of brain to body size grew several fold, and several fold again from primitive to modern mammals. This trend may be continuing slowly, as in the race between herbivores, which have to become more elusive or capable of defense, and carnivores, which have to become more wily. Primates, taking to an arboreal life some 30 million years ago, traded most of their olfactory capacity for more complex visual perception; and their grasping paws, able to manipulate objects, gave the brain more to think about.

Intelligence, however, complicates existence. As life becomes less mechanical, it becomes more burdened by problems of choice. The bacterium is programmed to do little more than grow and divide. A sea anemone, having chosen a spot to fix itself, has only to reach out to grasp passing plankton. The coyote faces far more numerous and complicated options—where and what to hunt, how to secure a mate, and so forth. Many animals seem to have an innate desire to explore;[17] they are playful and curious, seemingly seeking complications. For a human, doubts and vexing decisions, large and small, are almost the essence of life. The alternative is dullness and boredom.

As the organism faces ever more taxing choices, its apparatus for using information must be better geared to make responses agree with needs—to seek warmth when cold or coolness when overheated, chase prey when hungry, find a mate in season, and avoid danger always. Such responses are in part genetically programmed and unlearned, even in humans. So far as they are not inborn, they must be learned: one shrinks from very hot things by instinct, but painful experiences teach caution with stoves. Evolution does not so much dictate specific behavior as give capacities for it.[18]

For the intelligent animal, learning is mediated by rewards and penalties, pleasure and pain. It repeats pleasurable behavior and avoids pain-producing or potentially pain-producing actions; feelings play a major role in conduct. The sensations of pain and pleasure, or of discomfort and satisfaction, located in specific parts of the brain, are as important as sensory organs in the relation of the organism to the environment. Expanded and generalized into feelings of happiness and unhappiness, the pain-pleasure principle is the chief (but by no means the only) guide of human conduct, basically given by biological evolution, although skewed and redefined in a largely artificial environment.

## THE THEORY OF EVOLUTION

If the first key question of our existence is how the universe came to be, the second is how the multiform development of living creatures has occurred. Part of the answer is clear enough. The idea of evolution, the common origin of living things, probably going back to a single lucky molecule, is the foundation of biology. It brings together and makes understandable a host of facts. It makes sense of the fossil record, which is rich from more than 500 million years ago, when animals acquired shells or other readily preserved hard structures. Fossilization is irregular; it is not surprising, for example, that the origin of birds is poorly known, because flying animals are seldom preserved. But a glance at the remains of extinct creatures strongly suggests lines of development. Human ancestry is reasonably well documented back to clearly apelike creatures.[19]

Shared descent also accounts for homologous patterns. Thus bones in the feet of a frog or lizard, the paws of a cat, the digging tools of a mole, the wings of the bat or a bird, the flippers of the whale, and the hands of man all correspond, with modifications for different uses. Common ancestry must account for the fact that giraffes have the same number of neck vertebrae (7) as humans. Vestigial and greatly modified organs point to common descent, such as the remnants of limbs in some snakes, the token hips of whales, the pineal gland of mammals that derives from a reptilian third eye, or the ear-bones of mammals, shaped from reptilian jawbones. Embryology to some extent recapitulates ancestry; the human at one stage has gills and looks much like a larval fish. Changes come mostly as add-ons, because earlier changes occuring in the growth of organisms interfere more with subsequent differentiation.

The fact that species can be grouped into genera on the basis of shared characteristics, and genera into families, orders, and so forth, in the manner of a tree of ancestry, also indicates relatedness. There is a parallel in languages: the similarity of Romance languages points to common origin in Latin; resemblances between Latin, Greek, Slavic languages, Hindi, etc., point to the common Indo-European ancestor—a fact the discovery of which in the 18th century contributed to the gestation of the idea of biological evolution. Moreover, the way different forms are distributed on islands and continents corresponds well to the theory that they share ancestry and have come to their present habitats by virtue of geological changes and continental drift. This is most striking in study of the biogeography of such archipelagos as the Hawaiian and Galapagos Islands.

Common origins explain the fact that the basics are the same in all life, especially fundamentals of metabolism and reproduction by nucleic acids; the code relating nucleic acid bases to amino acids, although arbitrary, is invariant. It is most remarkable that the pattern of cilia, two central strands surround by nine, is constant from bacteria to humans.[20] But proteins differ in any species, even in all individuals except identical twins. The number of differences in the amino acids of corresponding proteins is roughly proportional to the degree of difference of forms—few between humans and apes, more between humans and carnivores, very many more between humans and jellyfish or molds.[21] Where the fossil record is incomplete, as it usually is, comparison of proteins is the most reliable method of estimating time elapsed since different forms separated from their common ancestor.

The idea of evolution, or common origins of creatures, was current, although not widely accepted, long before Darwin; and it is questioned only on religious, not scientific grounds. But it leaves open the other half of the key question: what brought about differentiation, change, and the development of new forms? It is Darwin's merit to have studied this closely and systematically and to have produced in mid-nineteenth century a plausible answer, which basically remains the accepted answer to this day. According to the simple theory, diversification and evolutionary progress depend on generally valid principles, which operate similarly in bacteria, fungi, trees, fleas, and monkeys. The basic fact is that organisms reproduce themselves, but do not always do so exactly. There are sure to be occasional variations, errors in the reproductive instructions, or mutations. Mutations, although random, may be increased by radiation or sundry chemicals; but their occurrence is unpredictable, if only because of quantum effects. A large majority of them are indifferent or harmful, but a few may be useful. If variations are inheritable, through succeeding generations some lineages will be better equipped than others to live and reproduce; and the proportion of the population with the favorable traits will increase, generation by generation, eventually altering the character of the population as a whole. Changes being cumulative over a very long time, organisms become gradually more complex and better able to cope with their environment in different ways. Isolated populations diverge to make different species in a million years or so.

Some sort of selection has to be a general principle, because all species must regularly produce more offspring than parents in order to continue to exist. Which individuals survive and reproduce is usually much more a matter of chance than of fitness; but inevitably some will succeed because they are somehow better equipped to secure food, escape being eaten, find mates, and reproduce. There have been numerous examples of natural adaptation to changed conditions in a brief time. For example, dozens of species of moths in regions made smoky by industrialization have become darker-colored and less conspicuous to birds in a few decades,[22] and vermin acquire immunity to pesticides with disconcerting rapidity. Animal and plant breeders have shown that by intensive selection (and hybridization) it is possible to bring about strikingly new forms not in millions of years but in centuries or decades—a fact that set Darwin thinking about differential survival even before his reading of Malthus suggested the survival of the fittest. Many human-engineered varieties, such as wheat, are very different from wild species; a few, such as maize, are quite new. Different breeds of dogs, chihuahuas, great danes, wolfhounds, and dozens of others, would unhesitatingly be classified as different species if found in the wild; the same is true of other domesticated animals, such as horses, cattle, and pigeons.

If artificial selection can bring striking changes so rapidly, it is easy to suppose that natural selection can accomplish much more over the ages. We can take it as a general rule that time's arrow converts unpredictable fluctuations into macroworld creativity through a mingling of order and freedom, stability and variation. In chemistry, when ingredients are mixed together and allowed to recombine, the more stable compounds remain; among living things, those best able to keep their pattern going by reproduction become dominant in their niche. In the realm of thought, more useful or more appealing ideas are carried forward and indefinitely reproduced, while others are forgotten. In the market economy, the ideology of which prevailed in Darwin's England as never before or since, competition assures that the better-managed enterprises will survive and grow.

There are many complications in practice. The reshuffling of genes through sexual process adds greatly to the possibilities of variation and adaptation; this is probably the reason that sex is nearly universal in all but the simplest organisms, although it is by no means clear why nature should so insist on it. There has also been random change in the genetic composition of populations, especially small ones, causing "drifting" independent of selective advantage. For this reason, island faunas are especially idiosyncratic. Oddly, the differentiation of species on the Galapagos Islands, which most of all inspired Darwin to formulate his theory of natural selection, is to a large extent ascribable to genetic drift, which rather contradicts the idea of adaptive selection. A great deal of variation in any event seems essentially meaningless. For example, there is endless variety of shapes of leaves in a single ecosystem under similar conditions, while leaves of a particular shape, for example pinnate, may be found in very different conditions, from moist shade to dry desert. There is little apparent relation between differences of human races and their environments. Darwin recognized this and sought an explanation in sexual selection, but it is not clear why there should be a preference, for example, for mates with straight blond or kinky black hair.

Essentially random events have also played an important part in evolution. Geography, changes of climate and habitats, and the separation or joining of populations have been very influential in the course of evolution, as one observes, for example, in the distinctiveness of the marsupial fauna of Australia. There have been many major episodes of extinction in the fossil record, when a large part, perhaps a majority, of previously prevalent genera and families disappeared, probably because of natural calamities, opening the way to new beginnings. Would dinosaurs still rule the earth if there had been no extraordinary alteration of the environment 65 million years ago?

With such qualifications, the branching of the tree of life and the progression from microbes to higher animals make intuitive sense in terms of adaptation through the survival of the fittest, or the natural selection of variations. For example, plants must have gradually become adapted to arid climates by reducing leaves to thorns and storing water in fleshy tissue under a waxy skin; cactus thus came into being. The cheetah lives by running down swift antelopes, so in any generation the fastest are best able to propagate themselves. Millions of years of such selection have produced running machines capable of breaking the highway speed limit. It makes intuitive sense—until one begins to inquire closely.

## PUZZLES OF EVOLUTION

The survival of the fittest appeals to the ordering mind. The Darwinian explanation is straightforward, simple, and prima facie sensible; and it can be stated in the impressive mathematical form of population genetics. The notion of a vital essence, or "vitalism," having been discredited, there is no testable or concrete alternative explanation; and those who doubt the theory of change through natural selection of random mutations are dismissed as simply unscientific. Biology textbooks give the impression that the problem has been solved; students are not to be bothered by doubts. For example, one states, "Natural selection, moreover, being as purely mechanical as gravity, is neither moral nor immoral."[23] Writers on evolution treat questioners with contempt; they seldom acknowledge basic doubts about the theory, although many and difficult questions have been raised. The problem of the evolution of instincts, for example, is much neglected.

The Darwinian approach is pleasingly reductionist, reducing causation to strictly material terms; and it is easy to believe that complete understanding awaits only adequate study. Many biologists uphold it with emphatic emotion because they wish to defend their discipline against mystic or religious claims that imply supernatural interference in the genesis of species, especially the human. They may admit doubts among themselves, but they will give nothing away to the creationist enemy.

Biologists differ most sharply regarding the tempo of change and the nature of transitions, which raise problems that cannot be swept aside. Transitional forms are rare. One stable species is succeeded by a related but distinct species without a continuum of forms; new orders appear as though from nowhere. The gaps are explained on the grounds that evolution is more rapid in small populations and intermediates are likely to be less abundant, unstable, and relatively short-lived before finding evolutionary success and stabilizing. This reasoning is plausible in regard to secondary changes, as from monkeys to apes. The linkage of major groups is much more problematic, however. A chart of the fossil record, showing the appearances and relative abundance of classes and phyla of animals or plants, shows only dotted lines of conjecture connecting origins. Most new classes appear full-blown without apparent predecessors.[24] The relative continuity from amphibians through reptiles to mammals is unique. A rare seemingly transitional form is the *Archaeopteryx*, which has some reptilian characteristics, including claws on the forelimbs, teeth, and a long tail; but *Archaeopteryx* is a clearly a bird, with wing feathers (a fantastic feat for natural selection) very like those of a modern bird.

The gap between a land animal and a whale is enormous, and there must have been a very large number of steps on the way; it is hard to imagine how they could all have been so uncommon as to fail to leave a fossil trace.[25] It is also suggestive that major classes are quite well defined, without intermediates with ancestral classes. For example, all birds have many characteristics that no non-birds possess, such as a unique respiratory system. In reptiles and mammals, air flows in and out as the lungs expand and contract; in birds, there is a much more efficient continuous throughput. Lungs do not expand, but air is pushed through tubes in them by means of a set of air sacs.[26] The two systems are so totally different that viable intermediate arrangements seem inconceivable.

It is most problematic to account for transitions. The whale's tail works up and down; its ancestors, as land animals, must have moved their tails sideways, as do most fish. Either direction is suitable for propulsion, but bones and muscles have to be structured to move most easily in one direction or the other, and there is no obvious way to make a gradual change or reason to do so. The sex of turtles is determined by the temperature at which eggs are incubated; it is not easy to imagine how there could be a shift from genetic determination by any useful stages. Countless more complex adaptations that the biologist takes for granted are difficult to understand in terms of the accumulation of small random mutations, even over a very long period. Innumerable structures or traits would seem to have no utility unless fully developed, and no one supposes that genes suddenly appear by chance to produce an entire organ.

It is difficult to avoid the conclusion that natural selection may fairly well account for microevolution, but the accumulation of rare beneficial mutations can hardly give an adequate explanation of the great leaps. To regard so simple a mechanism as the sole source of major innovations, new organs, and instincts is a statement of faith, not a view based on known fact. Biology faces countless unknowns, if not mysteries, as soon as it inquires how specific traits have come about. The genesis of very few organs is well established, and almost nothing is known about the origins of instincts.

It does not seem likely that a more complete fossil record will answer the major questions. A tremendous amount of digging since Darwin's day has turned up thousands of species, but the major groups remain almost as isolated as in the infancy of paleontology. Fossil birds have been found with some reptilian traits, and a very few mammals (monotremes of Australia) lay eggs and otherwise remind us of reptiles, but there is no scattering of holdover reptilian traits, such as might be expected if the group had slowly emerged from ancestral patterns.

Certain groups of animals seem to have special capacities for development. For example, the richness of instincts among insects, especially in certain orders, suggests that they have a greater facility in this direction than other animals—possibly a reason for the extraordinary success of the class *Insecta*, which comprises the majority of known species of animals. Various families of snakes have separately developed different poisons; evidently conversion of salivary glands to the production of venom comes easily to the order. It does not to mammals, useful as it would be for both offense and defense, a shrew with paralyzing saliva being the chief exception. Even-toed herbivores (artiodactyls), deer, antelope, cattle, sheep, and relatives, have paired horns of quite diverse materials and construction. Odd-toed herbivores (perissodactyls), horses and relatives, have no such horns. Unless certain mammalian potentialities go back to common ancestors, it is hard to account for the close parallels between Australian marsupials and very distantly related placentals elsewhere, parallels seen in at least a dozen types, such as wolf-like hunters, mole-like diggers, and jumping mice. The resemblances are too close to be attributed to accidents of adaptation.

Development often seems to proceed quite differently from what one would intuitively expect. For example, lines may evolve as though toward a predestined goal, perhaps several in parallel, as in the evolution of mammals from a reptilian branch.[27] Some forms, such as cockroaches, have changed very little across a hundred million years. It is hard to imagine that mutations in the cockroach genes ceased to occur and that there could be no improvements; moreover, the creature has remained fixed despite great changes in environment. Logically, evolution should proceed most rapidly in large populations, with many mutations and severe competition; the reverse is the case. The ability of small populations to diverge rapidly is remarkable; in some six million years honeycreepers in the Hawaiian Islands evolved 42 species; fruitflies, between 500 and 800, with all manner of habit from carnivorous to vegetarian.[28] It seems that the first pair of snails landing on the newly formed islands gave rise to about a thousand different species.

The sheer complexity of living creatures is baffling. The mechanism of the eye, for example, has puzzled many, including Darwin: the properly curved transparent cornea, the adjustable diaphragm of the iris, the automatically focusing lens, the millions of retinal cells, elaborately linked with one another and the brain, and the pigments sensitive not only to light but to specific colors. If Darwin had been aware of the fantastic complexity of vision at the molecular level[29] perhaps he would have given up. It is hard to believe that random changes in proteins produced by genes could produce this complexity even over a very long period.

Countless behavior patterns are quite as incredible. Sea slugs and some other animals eat cnidarians (jellyfish and relatives) as though purposely to acquire their stinging cells. The active parts of these are passed unactivated through the body of the sea slug and located in its skin to serve as a defense—seemingly an impossibly difficult adaptation to acquire by any random process.[30] Tiny brainless worms have "learned" to behave quite differently in successive hosts. One wormlet "instructs" the ant it parasitizes to climb a blade of grass, grasp it firmly in its mandibles, and wait during the hot part of the day when it would normally retreat to the nest, as though desirous of being eaten by a sheep, in whose liver a new generation of flukes will proliferate.[31] A complicating factor is that the ant is not likely to be a useful intermediate host unless it is made to put itself in position to be eaten by a sheep. Oddly, a fungus has a similar effect on ants it infects, directing them to make themselves into a platform for the dispersal of fungus spores. A trematode worm finds its way to a snail's tentacle, becomes much enlarged, takes on colors like an insect, and causes the snail to wander out during the day when snails like to hide, so that a bird will swoop down and pluck off the worm-filled tentacle.[32]

It cannot be by gradual stages that a worm adapted to life in the guts of a bird or mammal can acquire all the modifications needed to infect snails or other creatures to prepare themselves to get back to its principal host. Yet very many kinds of worms require one to three intermediate hosts and have as many as 5 larval stages. One fluke goes through a snail, a frog, and a small rodent to wind up in the mink in which it reproduces.[33] The prosaic ascaris worm, a common parasite in mammalian intestines, behaves ridiculously: eggs swallowed by the animal hatch in the intestine, burrow through the wall, enter the bloodstream, and emerge in the lungs; the small worms get themselves coughed up and swallowed, to lodge and grow to maturity in the intestines. How they could get started on such a journey is baffling. Once the wormlet is headed for the bloodstream, it is lost unless it is prepared to survive in a very different medium, find its way to the lungs, know enough to burrow out again, move high enough to be expelled, and then survive the corrosive juices of the stomach and find an intestinal domicile. Nothing seems to be gained by taking the adventurous trip to wind up at the starting point. Life does things the hard way.

Young golden plovers also make a remarkable journey. They linger at the nesting grounds in northern Canada after their parents have gone south and then follow by a route they have never seen to Patagonia, nine thousand miles away. Several species of birds migrate three thousand miles from Alaska to winter in Hawaii, where a small error is death. It is hard to imagine how this instinct could become established, because the islands have never been close to the mainland. Birds blown off course land accidentally on the islands from time to time, but there is no theoretical way such a detour could be converted into an inherited migratory instinct.

Birds and bees orient themselves by the sun, among other things; this requires coordination of senses with the needs of navigation in a complex fashion. This is remarkable enough, but some birds navigate by the stars—a fantastic set of instructions to establish in the chromosomes and the more difficult to lay down because of the fluxes: not only do positions of stars change by the hour and season, but the seasonal positions of constellations shift in the course of a few thousand years.

One can imagine woodpeckers gradually lengthening their bills because of the utility of such an instrument; but how two species of Galapagos finches could get started using twigs or cactus thorns to poke grubs out of holes is baffling, unless one credits them with a good deal of intelligence. A related bird pecks at sensitive spots on young boobies to draw blood to suck, a habit which likewise seems to imply learning.[34] A cliff swallow picks up an egg in its beak (with some difficulty) and deposits it in the nest of a neighbor. It thereby increases the number of its potential offspring by getting neighbors to raise its young. But unless this action is carried out completely, it means the loss of a good egg, and no one would suppose that a random mutation could establish such a complicated pattern. Did some bright bird think it up while bored with brooding?

Examples of baffling adaptations could be multiplied endlessly, such as the binary gas the bombardier beetle expels at near boiling temperatures, the electric organs of various fish, and the coded messages of the firefly. How such adaptations could have been brought about by random changes of proteins and enzymes defies understanding. But they are at least useful in finished form. Much behavior seems to lack logic; not only is it unclear how it could have arisen, but it has no apparent survival value. For example, the male bedbug, instead of mating in normal fashion, injects its sperm by puncturing the female's abdomen. The spermatozoa migrate through the body cavity to receptacles where they are stored until needed.[35] It is difficult or impossible to imagine any series of small, random modifications that could bring about such a gross change of procedure, requiring radical readjustments on the part of both male and female, with nothing apparently to be gained by this eccentricity. Some leeches have a parallel habit: spermatophores deposited on the female bore through the skin into the body cavity.[36]

For a more elegant example, monarch butterflies winter in a few small locations, mostly in central Mexico. In the spring they set out northward, but they go only 200 miles or so, lay eggs on milkweed plants, and die. A few weeks later, the next generation travels about the same distance, and so on for up to five generations to Canada. As the autumnal chill closes in, adults head south to the wintering grounds from which their great-great-great-grandparents started out, a very long and difficult trip for the slow and feebly flying insects. Not only is this a fantastic feat for the butterfly's minuscule brain, but it seems unprofitable. Much easier than the hazardous migration, during which a storm may be disastrous, would be for eggs, pupae, or adults to winter in the north, ready to enjoy the milkweed when it comes up in the spring, in the manner of other lepidoptera. Yet evolution commonly does the unobvious; and the monarchs are numerous, ergo successful.

Why should an entire class of ants (dorylines, or army ants) have lost eyes, although vision would obviously be very useful in their predatory life? What selective advantage could the female hyena get from fake male sex organs? Worse, some sea turtles likewise seem to lay only once in their lifetime; but they have obviously escaped the penalty of extinction. The same is true of some octopi. Most salmon also after spawning discard the parental bodies; king salmon and the related steel-headed trout save themselves to have a new family. Yet the suicidal fish are more abundant. Males of the marsupial mouse (antechinus) die soon after their first mating season;[37] it would be hard to argue that there is selective survival value in the early mortality of a nonseasonal mammal. With a different sort of perversity, the peach tree produces a toxin that prevents the growth of peach seedlings.

One would suppose that, of all traits, facility of reproduction would be most facilitated by the selective principle. This would often seem the case. Weeds succeed by virtue of tremendous numbers of seeds, and birds raise as many young as they can feed. Yet some animals, such as condors and rhinoceroses, have low fertility despite the most favorable conditions that humans can procure for them. Pandas have evolved a pseudo-thumb for stripping bamboo shoots, but where selection should operate most strongly and directly they fail: they mate reluctantly and produce unviable young. Evolution has made cheetahs marvelous runners, but poor reproducers. In view of scanty reproduction, a species limited to a few hundred or even a few thousand individuals may be in danger of becoming extinct despite the best efforts of friendly conservationists.

Very many animals reproduce much less abundantly than they seemingly could. Lionesses, for example, spend about 7/8 of their time resting or enjoying themselves and have cubs only at rather protracted intervals; their life, like that of most predators, is far from an all-out competition to leave the maximum number of descendants. It is as though they calculate not to overreproduce in order not to threaten the food supply of the whole population. But no such adaptation, even if one supposes random mutations could bring it about, would have survival value for an individual. It would have utility only for the entire population within an ecosystem, and even if established would be unstable because of the advantages for any small group with greater propensity to reproduction. Natural selection does not operate in favor of whole populations.

Evolution is an inconsistent mother. Chicks of the barnacle goose soon after hatching plunge off their cliffside nests on a journey to the water that costs the lives of about half of them; a more provident nature would surely have the parents bring them food until they are better prepared, or carry them, or endow the babies with means of surviving a fall of hundreds of feet. Why does the nature that teaches so many marvellous tricks not direct the tarantula either to run away from the attacking wasp or to fight back with its powerful fangs?

If the capricious goddess overseeing nature does many fantastic things, she ignores many simple opportunities. On the one hand, she creates the extravagant gaudy plumage of the bird of paradise. On the other, she denies to all animals above protozoa the enormously useful ability to digest cellulose, the chief component of plants. This would seem an easy faculty to achieve. It requires only an enzyme to split cellulose molecules into readily utilized starches, the sort of adaptation that should easily arise from the genetic mutations that natural selection postulates. Fungi, protozoa, and bacteria, after all, possess such an enzyme. Bacteria even seem able to quickly develop the means of digesting almost any organic material and even many synthetics. But termites and cows have to host protozoa to break down cellulose for them.

Looking at an animal such as the cow, one is struck by how suited it is to its way of life. It has broad teeth for grinding fodder, a facility for reprocessing it for more complete digestion (chewing the cud), four stomachs, and intestines modified to get the maximum from its low-calory, high bulk diet. The bovine plan clearly has survival value. Yet the horse gets along well with a simple stomach; and another herbivore, the gorilla, entirely lacks the cow's adaptations for a vegetarian diet. The gorilla's guts are very like the human's except for a larger stomach and a longer small intestine,[38] and it happily eats meat if offered. Why has the ape given up omnivorous habits for a vegetarianism for which it is not well equipped?

A different kind of adaptation is the placing of male gonads of nearly all placental mammals outside the body cavity. It can hardly be really necessary, as such animals as the elephant demonstrate; it is a considerable undertaking to make the shift; and these important organs become vulnerable. It would surely be easier to make some slight adjustment to permit spermatozoa to develop at ordinary body temperatures (if for some reason they had to develop a need for cooling in the first place). It is likewise puzzling why humans should have heavy menstruation, with wasteful shedding of the lining of the uterus, an inconvenient "adaptation" no other mammal finds necessary to a comparable degree. It is even more puzzling how the human female cycle has come to be geared to the sweat odor of males.

Evolution often seems to favor unnecessary complication. For example, a number of species of orchids make themselves so like female wasps—through scent, shape, and texture—that love-struck male wasps try desperately to mate with them and thus spread their pollen.[39] It would be simpler to produce a little nectar to reward the visits of bees, flies, beetles, or other insects than to contrive so elegantly to deceive a single species. Many birds go through elaborate and lengthy courtship ceremonies, the utility of which is not apparent. It would seem more expeditious for the couple simply to get on with the business of mating; the antics of their foreplay appear more like inherent playfulness. Why should it be advantageous for eels to grow up in fresh water and return to the ocean to spawn, while salmon do the reverse? In both cases, many remarkable adjustments are needed to move from one promising environment to another.

There are many contradictory behaviors, opposites that can hardly both be evolutionarily advantageous. For example, some animals will attack and kill strangers of the same species, while others with a similar way of life are pacific. Some mothers kill the offspring of fellow group members, while others, such as lionesses, suckle young of other members of the pride. Males of some kinds of monkeys kill young not their own; others are tolerant or affectionate. Males of many mammals are fiercely possessive of females; those of other species are indifferent to matings by presumptive rivals. Males usually contend for the female, but vixens fight for a mate. Many bird and some mammal fathers cooperate in raising the helpless young, as evolution would seem to dictate; some bird and most mammal fathers shirk this duty. Most birds build habitable to elaborate nests; others in similar habitats, for example fairy terns and peregrine falcons, make no nest at all. One can, of course, argue that either of the opposite actions is evolutionarily advantageous. The conclusion must be that very many traits or behaviors have no particular relation to survival value.

To have utility, adaptations ordinarily require various improbable involved changes to come together, none being of value without the others. Appropriate instincts are necessary to give utility to structures; for example, the rattles of the rattlesnake would be only an encumbrance without the instinct to use them as warning devices. There are complications even in such a straightforward adaptive trait as the change by subarctic animals from a brown summer to a white winter coat. This requires: 1) follicles capable of producing both white and brown hairs; 2) the ability of follicles to respond to a hormonal signal to cease growth of brown and begin that of white hair; 3) production of a hormone in the brain to cause growth of hair of a particular color; and 4) an appropriate connection of hormone-producing cells with the visual center to register shortening or lengthening of days. There are probably other complications. In any case, each of these changes by itself would be useless, and it would be astronomically improbable for them to come about jointly by any random process.

Some baby spiders are dispersed by climbing to a high perch, spinning a thread, and soaring away when a breeze catches it. But casual extrusion of silk can only be disadvantageous unless it is coupled with the instinct for using it as a sail, and it is not likely that the necessary instinct of itself could appear from a single mutation. Soldiers of many termite species shoot a poisonous fluid onto intruders. This requires: 1) production of the poison; 2) immunity to the poison; 3) a complex apparatus with tube and pump for projecting the poison; and 4) instincts for using it appropriately against enemies—all of these requiring extensive genetic instructions, none being useful without all the others.

Weaver ants bring leaves together to construct a well-engineered nest in tropical treetops. To do this, the workers stimulate larvae to secrete silk that the workers use to tie leaves together—an ability unrelated to the normal use of silk to enclose the pupa.[40] Must one assume that the ant builders are simply carrying out preset actions, somehow built into their minute brains? If humans devised such a method, we would certainly credit their intelligence. Much the same could be said of the archer fish, which knocks insects into the water by a jet several feet long. This requires a complex propulsive organ, exact aim, and vision corrected for seeing out of water, all coming together in a near-perfect form.

A female moth secretes a chemical, to which the male's antennae are phenomenally sensitive. The faculty of each sex is useless without that of the other, and it is not easy to see how the sexes could simultaneously make the appropriate adaptations by accidental process, or how the chemical signal, or pheromone, could become different for many different species, since it has to be very specific and any change in it would destroy its usefulness. Yet pheromones are universal among insects. They are even copied. The bola spider, which catches moths by swinging a sticky droplet on the end of a silk strand, emits the attractant by which female moths draw mates.[41]

Any slight deviation from the frogs' basic method of reproduction, laying eggs in or beside water, would almost certainly be harmful; yet they have devised many ingenious modes of giving their young a start in life. The female of a tiny tropical species carries newly hatched tadpoles on her back to a number of suitable pondlets, then remembers to return to them to deposit infertile eggs to feed the babies. An Australian frog even raises its young in its stomach. This trait can be of no great utility, because it is unique to a single rare species; but it requires extensive adaptations on the part of both frog and tadpoles; and there is no easily imaginable intermediate condition through which it could evolve. Any ordinary frog that began swallowing its young would immediately become extinct. Evolution has brought about such difficult means of getting over the disadvantages of laying eggs in water in lieu of something seemingly much simpler, such as the formation of a protective coating, as used by reptiles, snails, insects, and many other creatures, or internal hatching, which is rather common in fish and reptiles. Who or whatever decides such things seems often to prefer the imaginative way.

## QUESTIONS OF THEORY

If a simple organizing idea, natural selection, could fundamentally solve all the problems of evolution, this would be the only field of knowledge so blessed. No physicist would claim such a satisfying reduction of complexity to simplicity; the smallest things of which we have knowledge, electrons and quarks, behave in immensely complicated, indeed paradoxical ways. There is no simple idea ordering the facts of chemistry; the qualities of complex molecules have to be accepted for what they are, with whatever illumination may be gained from knowledge of component atoms. Much the same may be said of almost any branch of science.

Biology would also be the only field whose great secret was uncovered over a century ago. For lack of knowledge comparable to what the world now possesses, Freud was wrong in his ideas about dreams, and Marx's notions of economic development were erroneous. Darwin had no idea of the utter complexity of what he was trying to explain; it would be quite remarkable if, with the limited information of his time—most of his data he gathered himself—he hit on anything like an ultimate answer to the riddles of evolution.

The interpretation of evolution as the result of natural selection has clarity and definiteness. It rests on three simple and broadly acceptable axioms: that more offspring are produced than can in the long term survive, that there are random inheritable differences between offspring, and that those somehow best adapted survive in greater numbers and reproduce more of their own kind. It seems a satisfying answer that removes the need for further thought. Assuming the question is basically answered, biologists usually eschew unrewarding speculation as to how things could have come about and content themselves with reporting their discoveries.

But the attractiveness of the theory is disproportionate to the empirical evidence. It seems clear that variations, like the melanism of moths in industrialized regions, may come about by survival of the fittest, but there is no good proof that any species has been differentiated in just this way. Dogs have been extraordinarily varied by artificial selection, but no new species has been produced. Feral dogs around the world are much the same; different breeds revert to a general canine type.

There are facts about forests not derivable from the analysis of trees. To understand living creatures, one must study them as wholes, on their terms with their special patterns; and it is farfetched to suppose that any simple ordering principle can serve as a basis of explanation in the immensely complex living universe. Any reasonably explanatory theory will doubtless contain many subtleties; indeed, it may be that the diversity of living things requires a number of theoretical analyses regarding different groups and different kinds of problems. Principles of evolution may well differ in bacteria, mosses, birds, and primates. Insistence on a simplistic theory may impede the search for answers; for example, intelligent biologists fall into the absurdity of trying to account for homosexuality in terms of natural selection.

One should hesitate to postulate inexplicable forces. Interactions are complicated and creative; most genes have many interlocking effects, and many genes combine for one or many results. The human has about 110,000 genes, recombination of which through sexual processes permits endless experimentation. Chromosomes can also be altered in a variety of ways. The time has been very long, and many improbable things may happen in a few million years. Evolution is not subject to experiment in the ordinary sense, although close study can clarify many things; for example, research indicates that all vision has a common chemical basis and may have a single origin related to photosynthesis.[42]

Yet it seems clear that the idea of progress through the selection of beneficial mutations needs to be supplemented. Important adaptations must require several, perhaps very many successive mutations, each of which may be very improbable, and no adaptation is likely to succeed unless it has survival value at each stage or at least is not negative. Indeed, unless the survival value is large (which is not likely in case of a random change of a well-adapted species), it will probably be lost by accident; most lineages die out sooner or later. The difficulty that significant adaptations require concurrent changes in order to have utility is most apparent in connection with instincts, but it is a general principle. New organs need new muscular or neural connections, and so forth. Instincts are not a simple reflex, conceivably to be altered by a mistake in the instructions in DNA, but consist of a chain of actions with linkages and feedbacks; they could hardly be brought about by adding up the kind of steps envisaged by standard evolutionary doctrine. Yet the actions will be useful only if complete.

Chemists, like plants and animals, put together very complex substances by a step-by-step synthesis. But the chemist proceeds toward a goal, and only the final product needs to be useful. The living organism has to find its way to equally or more complex products supposedly without direction by numerous small steps, each of which has to have survival value. One might also compare fashioning the animal by the accumulation of mutations to making an automobile by a series of small changes made by error, under the impossible condition that each modification produce a more useful machine. Many things seem unattainable without a leap into a new form, just as a donut cannot be made from a sphere without breaking a surface. For example, the change from reptilian in-and-out breathing to the clever respiratory system of the birds requires more than mere improvements on the reptilian lung.

Such problems suggest that there may be more purposefulness in nature than biologists have usually been willing to concede. Many instincts would become less mysterious if it were supposed that animals, with their limited internal representation of the world, were occasionally able to devise useful actions, which are copied and become the inheritance of the species. Perhaps a bright or playful finch on a Galapagos island picked up a thorn and found it useful. A few years ago, British tits learned to open milk bottles, and the habit spread across the island.

Conceivably there is some means whereby learned responses become inheritable, permitting the development of otherwise incomprehensible instincts. Lions intelligently kill their prey by throttling them; one may postulate that this came about by selection of random changes of nucleic acids causing connections in the brain to be appropriately reorganized to coordinate perceptions of the struggling animal with necessary motions to clamp jaws on its throat. Or one can guess that lions, trying various means of killing, learned this effective strategy and assimilated it.

If there is purposefulness in the formation of instincts, this implies that behavior learned by the animal may somehow play a role in shaping its being. Thus the gorilla may be regarded as an ape that took to a diet of leaves and shoots instead of the omnivorous pickings of the chimpanzees, with anatomic adaptations lagging behind behavior. The giant panda likewise is anatomically an omnivore, like the bears to which it is related; for some reason, it gave up foods other than bamboo, which it is not well outfitted to digest. Like the gorilla, it retains an appetite for meat it normally never eats. If there has been some feedback from experience to heredity, the process has certainly not been so simple as proposed by Lamarck, who supposed that the giraffe lengthened its children's necks by stretching for high foliage. But was Lamarck partly right in that the giraffe began as an antelope that adopted habits of high browsing, elongation of the neck coming afterwards?[43]

It is a firm tenet of Darwinism that information flows only from genetic material to the body, never vice versa. This, however, seems questionable. All parts of the organism interact; and it would be surprising if there were no body-genes effects. Obviously, conditions in the body affect the expression of the genes in all multicellular animals; it is through feedback from the body to genetic instructions in particular cells that organs can be differentiated. Moreover, the hundreds of types of body cells to which the germ cells give rise breed true even when separated from the body; skin or liver cells, for example, continue to produce skin or liver cells when cultured in a test tube. That is, the invention of multicellularity required devising means to change, in effect, the hereditary instructions according to the role of the cell. This cannot have come easily, because it required nearly 3 billion years.

There have been reports of external conditions affecting heredity. If flax plants are grown extra large by special fertilizer, their descendants for some generations are large without benefit of fertilizer; likewise, flax plants dwarfed by being deprived of nutrients have smaller descendants through several generations. This also occurs with tobacco.[44] An acquired immunity may also be inherited.[45] A fungus parasitic on beans changes genetically to overcome the defenses of the host.[46] It should be noted that if at any time an organism somehow found means of adapting genetically to its environment, this would be highly advantageous.

It would be well to think in holistic terms: there are tendencies to pattern. The hearts and circulatory systems of birds and mammals are broadly similar and differ from the reptilian scheme in the same ways, although the common ancestor of birds and mammals was a very primitive reptile. The architecture of tetrapod forelimbs and hindlimbs is basically the same, for no obvious reason. Many animals have great restorative powers. For example, if the lens of the newt's eye is excised, a new one is grown from the iris (not from the skin, the original source of lens tissue).[47] Natural selection could hardly have brought this about.

A newt can hardly lose a lens without losing the eye, and no halfway lens would be of any use. But newts can regrow many lost parts, especially feet and limbs; there seems to be a general life-forwarding capacity to fill in gaps in the body pattern. This would seem akin to the ability of a part to become a whole, as when an early embryo divided makes twins.

Such matters raise the problem of morphogenesis in development. One should not expect deep understanding of evolution, which has left only a very partial record, until we understand the simpler question how undifferentiated eggs turn themselves into elaborate animals. There is likewise not much hope of understanding how incredible instincts have come about unless or until we know how they are coded and transmitted by nucleic acid—something that prima facie seems far more difficult than the formation of simpler bodily structures. To make a simple change of a computer's instructions, as from drawing a circle to drawing a square, may require a thousand bits of information; the simplest instinct would seem to need a number totally beyond the capacity of random mutation.

If exploratory behavior can lead to hereditary patterns, this would do much to explain countless otherwise inexplicable adaptations and to account for the phenomenal variety and inventiveness of life. Numerous adaptations suggest something like learning, in which selection would be a stabilizer or fixer of patterns but new integrations could not possibly have come about simply by mistakes in the genetic instructions. If mutations are like typos, it is easier to imagine them producing new words than new chapters. Natural selection is certainly a proofreader of the book of life, perhaps an editor; to give it credit for full authorship is too much.

One may or may not care to endow living creatures (as Grassé does)[48] with an "immanent finality," or share Koestler's view of a "creative force all along the lines toward an optimal realization of the creative forces of living matter."[49] But, in the words of a sober writer, living systems show "a whole new theoretical world, with a whole new physics associated with it." Furthermore, "We cannot in principle do biology within the confines of contemporary physics."[50]

Critics of Darwinism have no plausible alternative, but the lack of a plausible alternative does not prove a proposition true. The inventiveness of life seems too fantastic for any simple, chance-bound explanation; and if the laws of evolution are understandable, they are not yet understood. It is not unreasonable to see in living nature, as in physics, a manifestation of the penchant of the metacosmos for complexity. It is reasonable to guess that the patterns of life, while conforming to physical and chemical laws, go beyond them in the realization of the order-making capacities behind and in the universe, incorporating what we can fairly call a higher, or more creative level of being.

We may recall the anthropic principle. Our universe has to be such as to permit the emergence of beings capable of contemplating and studying it. For this, many utterly improbable physical parameters had to be precisely as they are. But this is true not only of the conditions for the formation of stars and hospitable planets. In biology as in physics things that seem unlikely or incredible may be necessary (for reasons much harder to elucidate than in physics) for the presence on this earth of beings capable of observing and commenting on their cosmos. There have to be ways of building the high complexity necessary for thinking creatures. Perhaps it is not possible to bring them about through the simple mechanism of variation and selection. If this is the case, there must be subtleties to be discovered in evolution.

We learn something of the nature of the cosmos and metacosmos from the mysterious ways of physics, and we should learn something of the nature of being and ourselves from the marvels of living things.

# 3

# THE CRUX OF CIVILIZATION

## INTELLIGENCE

The third miracle, following the birth of the cosmos and the genesis of self-reproductive and self-improving life, was the emergence of self-aware, critical, and self-compounding intelligence in society, or civilization. This, also, like the dance of the honeybee and the navigation system that directs the young tern nearly halfway around the world, remains mysterious.

Understanding the evolution of the human capacity for abstract reasoning is the more difficult because the growth of the brain occurred under conditions of which little is known; and social relations, which leave no fossils, were doubtless important. Many unweighable factors probably played a part. For example, among animals, sexual selection has promoted many traits, such as huge antlers, which are burdensome and of little utility except for jousting. Perhaps near-human nubile females preferred clever suitors, and the craftier or more eloquent or more capable had more numerous offspring or could feed them better from a larger portion of the hunt.

Humans doubtless were also driven to become smarter because they faced the most challenging adversaries, other humans, not in single combat like stags or seals, but as groups capable of planning strategies. The cunning devoted to snaring a tiger was equally or more necessary in contending with malicious neighbors. Clashes between bands, perhaps mostly over hunting grounds, were inevitable, just as they have been chronic among primitive peoples of modern times; and the ability to scheme and deceive must have often meant the difference between victory and defeat, when victory gave prosperity and multiplication, and defeat brought deprivation, if not death. There is no limit to the intelligence useful in conflict.

It is not necessary to suppose that the brains of paleolithic humans functioned in quite the same ways as those of moderns. A complicated apparatus, such as a computer, turns out to have capacities unplanned by the builders; and much of the power of the human brain lies in its extreme flexibility. The brain developed as a sort of general purpose computer, and as such it has an indefinite range of utility in an unlimited variety of applications. The ability of an African herdsman to recognize individually a thousand cattle of a single breed—an ability of the sort that would have been useful to our distant forebears—may transfer to more modern mental feats, such as memorizing tens of thousands of verses of Hindu epics or remembering the characteristics of thousands of chemicals.

We find it difficult to account not so much for the ability of the brain to handle data, in which it is far outclassed by a modest computer, as for its capacity for generalization and abstraction. It is not easy to understand how evolutionary selection among primitive hunter-gatherers could have brought about an organ capable of composing symphonies, or inventing recondite mathematical theorems. Nor is there an explanation why for some individuals there is no greater pleasure than the expansion of the mind and its exercise, be it in devising a subtle strategy in chess or gaining a new insight through a novel mathematical theorem.

Intelligence outruns apparent utility in some nonhuman creatures also. The gorilla, for example, is herbivorous, pacific, and without predators; it would not seem to have much need for an excellent brain either to find food, fight, or escape danger. Yet it is much brighter than mammals such as gazelles, which have more obvious need for wits to stay alive. Elephants are quite intelligent, although intelligence would not seem to be more necessary or useful for them than for sheep. For the lumbering beast, as for the big herbivorous dinosaurs, a pygmy brain should be ample. But with its size, the elephant acquired a large skull, a brain several times bigger than the human, and impressive mental abilities.

The case of the cetaceans, dolphins and whales, is still more puzzling, even if their brains (sometimes far larger than those of humans and even comparable, in the case of dolphins, as a proportion of body weight) do not quite measure up to the claims of admirers. The aquatic mammals are much farther from being able to use tools or to take complicated actions than elephants: flippers are for swimming only, and their watery environment is less varied and less rewarding of thinking. They have little more need for complex behavior than the small-brained sharks with which they compete, and less cooperation is involved in hunting fish or straining plankton from the seas than in lions' hunting zebras. Only man troubles their tranquillity. Their principal intellectual problem would seem to be boredom. Yet, possibly to escape this, whales, young and adult, play a good deal with the very limited materials at hand, such as floating logs.[1] They easily learn a great variety of tricks quite unrelated to oceanic life, and the songs of some whales suggest a lyric spirit—although biologists prefer to suppose they have to do with sexual selection. It may be that large networks lend themselves to complex information processing and perhaps something like consciousness, but why should evolution lavish such large networks on them with so little need? Their intelligence is at least as mysterious as that of humans and as difficult to account for in strict Darwinian terms.

## BECOMING HUMAN

Chimpanzees and gorillas are about 99% human; that is, they are said to differ from us in only about 1% of their/our genes and the proteins they produce. We have no types of cells not found in them; the few differing genes chiefly control the relative growth of body parts.[2] Like us, the great apes vary markedly in physique, temperament, social ways, and intelligence; and they behave rather like uninhibited humans. They classify objects and perceive analogies, for a time learn sign language as fast as children learn speech, have about a dozen different calls, cooperate in hunting, and have better locational memory than most people.[3] Chimps will work for tokens with which to buy bananas. Like humans and unlike lower animals, they have to learn such essentials as mating and caring for their young.

However, the great apes, unlike monkeys and baboons, have had limited evolutionary success despite their surprising mental powers. They have restricted ranges and small numbers even where they do not compete with humans. Their evolution seems to have come to a halt when poised for a takeoff with large capacities for information gathering, processing, and response. These assets were not sufficient for the unfolding of higher culture, and it would not appear that their way of life has significantly changed for millions of years.

The failure of the chimpanzees apparently arises from a low degree of socialization. With a brain of roughly the same size as early hominids, chimpanzees can develop and pass on techniques. For example, some bands know how to crack nuts with stones, but neighboring bands may lack this simple and useful art.[4] There is little transmission of inventions in feebly structured chimpanzee society. Families are fatherless, and there is continual violent contestation for status. Adults sometimes even cannibalize the babies of fellow group members.[5] Bands fight neighboring bands and kill their males.[6]

Better social organization seems to have been a prime requisite for the beginnings of humanity as it has been for our success ever since: a well-structured society and a fair degree of coordination of individual purposes makes possible learning and productivity. Perhaps the key advance came with the first regular use of weapons: as soon as a weakling had a spear, he could kill by stealth, and dominance required leadership and consent. Status could no longer rest on mere strength or size.

The emergence of protohumans from the static conditions of the apes around 5 million years ago is commonly associated with climatic change and a shift from jungle or near-jungle to bushlands, savannah, or grasslands. These offered a much more abundant potential food supply in the shape of large grazing animals; but life away from the safety of trees necessitated mutual protection.[7] Prehumans faced large predators more than twice as swift and far better equipped to kill with fangs and claws. But a dozen near-men with crude spears could easily cope with a lion. Our ancestors crossed a threshold when they began bringing down beasts larger than themselves and butchering the carcass for the benefit of the group.

The bonding of individuals improved with upright posture, which freed the hands, favored the use of tools and weapons, and also made it easier to pack parts of the kill back to kin and family. This preceded any considerable enlargement of the brain;[8] *Australopithecus* began walking erect more than three million years ago. It was also much more practical for the erect prehuman than for the knuckle-walking ape to carry young children.[9] Only humans hold infants in their arms for more than a day or two after birth.[10]

In the emerging new way of life, the brain began its expansion to its present size, the fastest evolution of any organ, so far as known, ever to have occurred.[11] But the near-ape only slowly grew into human ways; and the outlook was probably uncertain until culture reached a critical stage of self-reinforcement, like the spontaneous combustion of a pile of oily rags. *Homo habilis* made crude pebble tools 2.5 million years ago or more,[12] when the brain was only 20% larger than a chimpanzee's; but they were hardly improved over hundreds of thousands of years. Perhaps a million years ago our ancestors were using fire.[13] The rate of growth of the brain peaked between 500,000 and 300,000 B.C. There must have been circular causation: a larger brain enabled the humanoid to make better tools and use them more effectively in cooperation, which made tools and intelligence more useful. The shared hearth helped to integrate groups, and social interaction raised ever-increasing complexities. Large prey could be more effectively utilized by a larger group, and cooperative hunting and living made communication more useful and necessary.

Estrus was lost, love was born. Sexuality among apes is temporarily stimulated by pheromones and anatomical displays, such as colored skin patches; among humans, it is lastingly evoked by social interaction.[14] Nonreproductive sexuality is peculiarly human. Pair bonding and fatherhood followed; family life became advantageous because offspring helpless for many years needed paternal support.[15] Without it, the mother could have children only at long intervals. With the support of their mates, incidentally, women could bear young at shorter intervals than female apes and so crowd out their less progressive neighbors.

The father's sharing in the raising of young also doubled the transmission of culture. The prolonged dependence of the young, which facilitated the accumulation and transmission of stored information or culture, made individual skills and specialization possible. It is uniquely human to spend so much time learning about the world. A dozen years of immaturity, which would be overly burdensome for a single female or even a couple, became the means for the human group to improve its skills and tools.

Sharing of food and shelter made possible a close-knit social existence, probably in bands of 20 to 100, like those of recent primitive-hunter gatherers.[16]

Language became the prime distinction of the new masters of the plains. Those who have raised chimpanzee babies have felt very frustrated that the obviously bright youngsters, who could learn signs easily, could not be brought to produce more than the most rudimentary speech sounds. The speech centers of the brain are different from those controlling vocalization in nonhuman primates. Mentally defective persons may be able to speak fairly well, although they may be less intelligent than chimpanzees in other ways; the propensity to speech is inborn. The fundamentals of language are similar in all peoples, and pidgin languages developed for communication across cultures show parallels around the world.

Symbolic representation in infinite variety by means of language makes it possible to generalize experiences, form concepts, manipulate them mentally, and transmit them. It not only communicates meaning but creates meaning and expands intelligence. It facilitates the accumulation of information, specialization, skills, and hence more learning, which raises capacities and the power and utility of language. It is also necessary for a better social order, favoring altruistic and cooperative behavior, by which intelligence becomes more useful.

Dependence on tools and cooperative living affected physique. Teeth became weaker as long as 4 million years ago,[17] evidence of the decreased importance of physical combat; further decline was associated with the use of fire to make food easier to chew.[18] Muscular strength decreased. Gorillas are several times as strong as humans of equal weight, and chimpanzees can easily kill humans much larger than themselves. The loss, for reasons unclear, of body hair necessitated or was made possible by the use of fire and clothing. Humans became highly specialized in dependence on a single hypertrophied organ, the brain.

This specialization, however, made possible a higher degree of generalization, that is, adaptation to a very wide range of conditions. Humans could not only hunt all manner of animals but also gather a great variety of foods. With skins for clothing (made possible by good knives), along with fire and housing, they ranged to climates too harsh for naked tropical creatures.

They gradually multiplied, spreading over Eurasia by 50,000 years ago, staking out almost all the habitable regions of the globe, and becoming cosmopolitan-dominant long before the dawn of history. In the voyages of discovery of the fifteenth to the seventeenth centuries, the creature European explorers met almost everywhere, from polar ice to Pacific atolls, was their own species.

## PRE-CIVILIZATION

For hundreds of thousands of years after 500,000 B.C., cultural progress hardly reflected the expansion of the brain and presumed improvement of intelligence. Millennium after millennium, very little more was added than was forgotten. There must have been discoveries and inventions. A paleolithic Newton may have observed how the pattern of light and dark on a round stone in the sunshine changed according to the angle from which it was viewed and guessed that the moon was a sunlit sphere. But the idea, meaning little, would have raised no notice. Or someone may have picked up shiny reddish beads of copper that appeared when greenish rocks were heated in a charcoal fire, but no one cared.

Why innovation has prospered in certain places and times and stagnated at others is obscure in the historical record; it is more difficult to speculate how and why prehistoric culture inched forward. Growth of culture is far from automatic; it is suggestive that Australian aborigines were making only about 40 different articles when Europeans arrived,[19] perhaps no more, certainly not many more than when they settled on the continent 40,000 years earlier. Perhaps growing populations and larger societies encouraged new ideas, and wider communications stimulated improvements. Climatic changes required new ways of life. Needs of war doubtless played a part, as they did in historic times.

During the long cultural stagnation, however, some important inventions spread widely and became permanent. For example, around 350,000 B.C. an improved hand ax was carried rapidly over Africa and much of Europe and Asia.[20] In what seems a paradox but probably has profound significance, average brain size ceased to increase about 100,000 years ago (and has since then decreased slightly) just as cultural factors were coming much more strongly to dominate human existence. But the irregular accumulation of knowledge and techniques slowly raised the species above its simpler past without further apparent improvement of biological endowment. About 60,000 years ago a young man in Iraq was buried with eight kinds of flowers, as shown by pollen grains; evidently a new awareness and social feeling had entered human consciousness.

About 35,000 years ago, the more primitive but big-brained Neanderthalers were replaced by Cro-Magnons, or modern humans, who were physically weaker but culturally more powerful.[21] Their superiority was possibly due to better development of the speech center[22] and consequently improved social structures. The invaders, who had been living for thousands of years in the Near East before displacing the Neanderthal people, brought an upsurge of creativity. Many innovations appeared, such as sharper knives, harpoons, bone needles to make skin garments, shell necklaces, and Venus figurines. There were bone flutes and whistles, which must have lightened winter evenings. Beautiful cave paintings bespoke a new vision.[23] Stone age hunters, equipped with flint-tipped spears, then arrows, fashioning tools of hardened wood or bone as well as stone, made themselves masters of the earth, even exterminating many large mammals in the first human-caused ecological disaster. They provided their dead with ornaments, made cult images, cared for the disabled, and presumably led an interesting and rewarding existence. There seem to have been social networks of some kind over thousands of square miles; as in human inceptions, wider interaction must have been a key to progress.[24]

About 12,000 years ago, such inventions as pottery, weaving, and the domestication of such animals as dogs, goats, and sheep enabled more people to live better. Learning not merely to gather but to plant for a distant harvest brought a revolution like that of learning to hunt large animals long before. Agriculture multiplied the food supply and consequently the population ten to a hundred fold. By bringing many more people into settled communities, it permitted more specialization and more complex interactions whereby individuals could communicate and join their abilities.

Neolithic villages gradually grew into towns and what we call civilization. Trade must have played a part. Some places had supplies of superior flint, ornamental shells, or other valuables; and commerce concentrated wealth and population and widened horizons. Cults also may have contributed, as theocratic centers became meccas for those who wished the support of a god and his priesthood. Possibly religious persuasion was mingled with conquest. As populations grew and villages came into conflict, the idea of depriving neighbors of their crops, livestock, and perhaps women was inevitable. After pillage, subjection is a natural step; and a small settlement could grow, fortify itself, and build temples and primitive palaces by forcibly extracting part of the product of neighboring peoples. Well before the opening of the historical record there probably were small empires, perhaps fairly large ones. The neolithic empire of the Incas imposed a sophisticated despotism extending two thousand miles along the Andes.

## CIVILIZATION

Some ten thousand years ago, agricultural peoples of the Near East were becoming increasingly specialized producers of beads, pottery, stone tools, and many other goods, and had begun to work copper. They lived in towns, sometimes walled, of several thousand inhabitants, and evidently had complex stratified societies. As towns became cities, new needs in a more complicated existence brought a more involved social order, with new occupations and hierarchic institutions. Writing, an invention second only to language in culture-building significance, began with marks that recorded quantities of goods, probably to keep accounts of dues to king or temple. By giving permanence to information and enabling messages to be sent reliably from place to place, writing (at first on clay tablets) made possible an explosive growth of knowledge. It raised intelligence to a new mode of self-enlargement through the linkage of many minds, past and present.

The Sumerians, who invented writing about 3500 B.C., pioneered the new era. They were irrigation farmers in the lower Tigris-Euphrates valley, organized into a dozen or so autonomous city-states. They made a host of basic inventions, such as sailboats, baked bricks, wheeled carts, the potter's wheel, cylinder seals, the ox-pulled plow, and ornate metal ware. They built huge temples, composed literary epics, had formal schools, and created fairly elaborate political institutions.

Within about a thousand years after the Sumerians led the way, many other civilizations sprang up, generally following the same basic patterns. Of these, the Egypt of the pharaohs is best known, mostly because the dry climate preserved its artifacts. Others arose in the Aegean area, Western India, China, Southeast Asia, Central America, and Peru, all apparently closely or distantly inspired by the Sumerian way of life.

The growth of civilization, however, has never been automatic and self-sustaining but spasmodic and always contingent on socio-political organization. About 2300 B.C. Sargon of Akkad conquered the Sumerian homeland and founded the first empire known to history, which claimed to rule the world. As one empire succeeded another in the region, rationalism declined, progress virtually ceased, and for almost two thousand years there was more impoverishment than enrichment.

Such has been, broadly speaking, the course of history. A group of loosely affiliated competitive states has opened the way to invention and innovation by its stimulating atmosphere, fluidity, and good social order. Eventually, in every case except (thus far) our own, the free sovereignties have been overtaken by an extensive empire. Rulership standing for fixity and order, the priority of domination over cooperation, has brought creativity to or near a halt.

This has occurred in many variations, as in Egypt, India, China, Peru, and the Graeco-Roman world. The Egypt the classic Greeks knew abhorred change and had long since come to a cultural dead-end. India was an intellectual garden about 500 B.C.; thereafter it was for the most part a desert until forcibly brought into the sphere of the modern world. In the rigid hierarchy of the Inca empire, there was no room for experimentation; in the same territory a thousand years earlier, the variety of pottery suggested a lively contest of ideas. The New World civilizations of both Mexico and Peru early reached a plateau roughly equivalent to that of Egypt of early Old Kingdom times;

and when Spanish explorers found and overthrew them, they had not improved their arts for many centuries.

There is no reason to suppose that any of these static cultures, if left to itself for millennia, would have attained a breakthrough to electronic civilization. On the contrary, the tendency from century to century seems to have been usually toward more petrified societies. Old civilizations become ever more bound to the ways of the ancestors, more ritualistic, more superstitious, more apathetic, and more dominated by the apparatus of rule and parasitic classes, landlords, priesthood and officialdom. At the same time, population probably presses on the means of subsistence, perhaps (as in the Near East) lowering the fertility of the land.[25] Stagnation comes much more easily to human society than change and improvement.

There have been some ten or twelve autonomous primary civilizations—Sumerian, Egyptian, Indus Valley, Mexico, Peru, Shang China, Minoan, and a few others of comparable technological level. But they were all submerged by conquest. In three areas—Greece, India, and China—civilization resurged to what might be called classical sophistication. Again, in each of these, the lively culture was choked after a few centuries by imperial despotism. In one case, there was a rebirth of creativity: after the wreckage of the rotted Roman empire had been shattered by Germanic invasions, a new upsurge slowly developed, giving rise to the self-conscious science of the West.

The development of culture and technology, in other words, does not represent simply an accretion of knowledge but requires an uncommon social and political order, without which the potentialities of the mind are negated and there is no progress. But we have even less understanding of why or how cultures and civilizations arise and flourish than we do why animal species take their different courses.

In some societies there are incentives for independent thought, discovery, and innovation, with openness and unpredictability instead of conformity, and not only capacity for but receptivity to invention. The most prominent condition is clear: civilization-building has progressed chiefly in groups of independent states, actively competing, sometimes fighting, but sharing in trade and culture.[26] As at all levels, from molecules to nations, moderated disorder gives rise to new order: there must be sufficient freedom and incentives to encourage intelligence—like a random input—along with adequate organization to utilize and reward it.

## RISE OF THE WEST

The most familiar example of the sequence from competitive sovereignties to universal empire is that from the effervescent competitive world of the numerous classical Greek polities to the ultimately static Roman domain. The Greek city-states grew up in the centuries before 700 B.C. on the islands and peninsulas of the Aegean Sea on the fringes of Near Eastern civilization. Hundreds of statelets traded with one another, sharing culture and deities; they warred frequently but usually not destructively and rivalled keenly in sports and arts. Few in numbers and without important natural resources, excelling only in intelligence and morale, they enriched themselves and the world not only with literature, science, art, and philosophy but also by improved ways of doing and making things, from statecraft to shields and pots.

For several centuries, Greek civilization was copied and admired widely across the Mediterranean world; Greek literature and trade had no rivals. But if larger states paid symbolic tribute to their cultural masters, they were more interested in military strength. Macedon in particular assimilated enough of the Greek achievement to crush Greek independence in the 4th century B.C., and went on to conquer a huge empire. After the early demise of Alexander the Great, however, his domain broke up into medium-sized sovereignties, such as Macedon, Syria, Egypt, Rhodes, and Judea, along with many leftover city-states. This less divided and more imperial Hellenistic world was less original and vibrant than classic Greece; but it was highly productive, especially in natural science.

The master empire-builders of Rome vanquished the Hellenistic kingdoms, copied their art, and brought their science to an end—not because science was prohibited, but because the spirit of free inquiry passed with the spirit of freedom, and individualism and science ceased to be rewarding. Greek philosophers inquired into the nature of things; philosophers of the Roman empire asked how best to live within the regnant order.

Technology was retained mostly so far as it was useful, especially for the monumental constructions in which the empire took pride; but little was added even to architecture. Education declined to formalism and study of the classics. Astrology, a legacy of the Near East which Greek rationality had rejected, revived and replaced astronomy. As generation succeeded generation, there was less and less innovation; there was no feeling that anything could or should be new. During some six centuries, science slumbered, and even practical learning decayed. Literacy, which the classic Greeks took for granted, became a rarity in the Dark Ages after the fall of Rome. Science did not reattain its Hellenistic level until some 1600 years later.

When the Roman empire broke up, however, society was loosened and recovered the capacity for change and discovery. Even in times of widespread disorder, simple practical improvements were made or adopted, such as the stirrup, the horse collar, and the wheeled plow. As early as the eighth century the West was gaining ascendancy over Asia in military technology with better armor, the crossbow, and the like—an ascendancy which saved it from external conquest.

With the stirring of trade and town life from the 9th and 10th centuries, hundreds of political units across Europe were able to assert more or less autonomy, from kingdoms and duchies to bishoprics, and, above all, a host of free or semi-free towns. For 500 years, from the mid-10th to the mid-15th centuries, the industry, trade, and arts of Europe were dominated by the towns, usually called "communes." There were many constellations of them, in northern France, southern France, eastern Spain, the Low Countries, western Germany, the Baltic region, and Italy. They owed much to the historical accident of the Holy Roman Empire, a would-be universal sovereign without sufficient strength to be oppressive. The Italians were especially favored by the protection of the papacy, which was perennially at odds with the Empire. Their city-states were the freest, lasted longest, and prospered best; in Renaissance times they led Europe economically and intellectually. Many of Shakespeare's plots were laid in Italy, and Milton wrote much Italian verse.

But material progress and inventions, such as artillery, enabled monarchs to subdue towns and feudal vassals, forming centralized polities. After mid-fifteenth century the rising nation-states—France, England, Spain, Portugal, Austria, and others—became the masters of Europe. However, unification halted at this level, despite enormous further improvements in means of warfare, transportation, and administration. It is unlikely that Western Europe, if unified by some would-be conqueror—there have been many candidates—would have escaped imperial stultification. But an open international system during five centuries permitted the building of modern civilization.

This was extraordinary good fortune. Probably the most important single factor in it has been the geographic dividedness of Europe, like the dividedness of the islands and peninsulas of Greece. Mountains, especially the Alps and Pyrenees, form many natural boundaries. The peninsular character of Italy and the insular position of Sweden have been important. Without Swedish intervention, the anti-imperialists would have surely lost the Thirty Years War, and national independence might have come to an end. Most crucial has been the channel-moat protecting Britain. After England lost its continental holdings in the 16th century, it adopted an unswerving policy of protecting its independence by preventing the unification of the continent facing it. Without ambitions for continental territories, England fought many wars to sustain the balance of power, frustrating the ambitions of Louis XIV, Napoleon, and their predecessors and successors. The bogs of Holland made possible its independence; and the Netherlands, like England, was a powerful defender of the freedom of nations. In Eastern Europe, where natural divisions are much less formidable, nation-states have been unstable, multinational empires have prevailed, political freedom has been meager, and science and technology have lagged.

Cultural factors also reinforced the selfhood of the nation-states. The Reformation, in large part an affirmation of the independence of nations, triumphed where supported by the state; and its effect both in Protestant and in Catholic Europe was to affirm the power and legitimacy of secular authorities. The development of separate national languages provided a basis for political separation without destroying the essential cultural community of Europe. From the beginning of the sixteenth century, overseas colonization distracted the major powers from territorial ambitions in Europe—dubious wars for a border province were less attractive when huge expanses were to be had at little cost beyond the seas.

The rise of the nation-states in the 15th century coincided with the invention of printing, the third cultural revolution after language and writing. Printing permitted the accurate but inexpensive multiplication of information on a scale previously undreamed, not only ideas but all manner of practical instruction manuals. It made possible a signal upsurge of objective learning, leading to the high science of the 17th century and the industrial revolution that began in the 18th.

This could not have advanced rapidly had not Britain, one of the most progressive and commercially-minded powers, possessed good deposits of coal and iron within easy access of each other and water transportation. This fortuity permitted the development of a large-scale iron and steel industry—charcoal was a limited resource, and no heavy industry could afford to transport coal and iron ore more than a few miles by animal cart. The industrial revolution in turn made possible the construction of railroads, which facilitated the industrialization of the continent.

Industry, built on science and technology, brought Britain power and prosperity; and the expanding Western world turned to the study of nature with increasing eagerness and success. Chemistry and electricity brought industrialization to a much more sophisticated and productive level. Finally, in our day the computer has, like writing and printing, opened a new era in the generation, spread, and utilization of knowledge. This has brought a new stage in the growth of the information-sphere of the cosmos, a new kind of being (Karl Popper's World Three), standing far over not only the animal world but over individual human thought.

## Problems of civilization

Our technological civilization represents an advance over the ways of the apes comparable to that of the living over the nonliving, a novelty in the history of the earth—or of the universe? The life of individual organisms is no longer shaped by their physical or even mental capabilities so much as by artifacts and ideas; and culture is not even the possession of the individual and groups but is made by countless minds and belongs to the millions. The principles of genetic selection have ceased to mold the species, and its biological destiny is uncertain. If conditions of life formerly changed only through hundreds of thousands or millions of years, the way we work and live is now transformed in decades.

For the last thousand years, each century has found technology and learning at least a little stronger than the century before. But the forward thrust was somewhat hesitant and episodic for many centuries, and probably insecure, as periods of stagnation or regression succeeded times of cultural upsurge. Modern science is an unlikely product even of a fairly advanced civilization; there have been many potential cutoffs on the way. Only since the 15th century has movement steadily accelerated, leading to the recent explosion of knowledge of the world and ways of doing things.

We take for granted that science is useful and that organized intelligence can serve our needs and advance ever more powerfully; but the wellsprings of progress may be running out. We march ever faster, in a microsecond of geological time, into a strange new environment of new capacities. We stand at the pinnacle not only of power over our surroundings but of rapidity of change and instability, that is, insecurity.

The most obvious material problems, such as the nuclear danger, degradation of natural resources and environment, and management of population numbers, are technically solvable if there is a will to solve them. But there are subtler and deeper problems. They are questions of order and motivation, of reconciling individual and social needs and demands, of making technology really serve human needs, and of keeping lively the search for truth.

The transformation of the human way of life has come about less because of any particular device, even so revolutionary as the printing press, than from the discovery of the scientific method and the incorporation of systematic discovery into the social fabric. This requires open-mindedness, inquisitiveness, incentives for discovering new facts apart from their utility to particular persons, and willingness to test and apply different ideas. If the scientific-rationalistic approach were abandoned, all the pomp of civilization would wither away; if knowledge of things were lost but the urge to question nature and build up systematic understanding were retained, all the rest could be soon rebuilt.

It is questionable whether humanity can maintain appropriate but ever-changing institutions to assure openness and the ability to generate and use new knowledge. Contradictory ends must be served. It is increasingly necessary to cope with problems on both a long-term basis and a global scale, in ways sure to seem injurious to many. At the same time, society must avoid giving excessive powers to holders of political and economic advantages. Effective collective action is required, yet the freedom of nations is necessary, and freely competing powers may well be mutually frustrating if not destructive. Government must be strong; but irresponsible government can subvert the purposes of the community in the future as it has in many places in the past, choking invention and innovation and bringing stagnation and decline.

The problem of power, a secondary need of the personality, becomes the more acute as primary needs are filled. It has never been easy to maintain sufficient group solidarity for good order along with enough individualism and freedom to encourage effort and achievement. It is the more difficult as political power acquires more tools of coercion and deception. Modern civilization provides means for free expression and individual and group autonomy; it also provides means of manipulation, control of opinion, and repression.

If it is difficult for individuals to set aside natural impulses and meet the artificial needs of civilized society, it is far more difficult for a society to rearrange itself. We are cells of a superorganism, the more utterly dependent as civilization rises higher; yet we are self-centered and self-motivated. Society cannot prosper unless individuals act in many ways on their own, even contrary to the expectations of society, just as it cannot prosper if they act entirely on their own.

"Society" consists of individuals, a few of whom are placed or place themselves over the whole. The fundamental questions of social organization are how it is to be decided who should hold the power necessary for the coordination of the complex whole and how much those in positions of authority can rightly command—the relation of personal authority to the entirety. The choice of leadership becomes more crucial as the state undertakes ever more tasks; yet there is no professional expertise in statecraft corresponding to the mental equipment of the engineer, and no satisfactory way to elevate the most qualified.

This problem of government is deepened by inequality. There have always been lords and serfs, rich and poor; in the modernizing society, there is the inequality of the modern, that is, highly prepared, over the unmodern or unqualified. A few are far in the lead, adding to the capacities of society; some profit grandly from the complicated potentials of the new age; many carry part of the burden; others are merely passengers; some are quite left behind. This is very different from, and less healthy than the primitive society, in which everyone knew nearly everything important to know (except perhaps some spells of the shaman) and all shared in the labors and rewards of the group.

The human condition is also at risk because only about a third of the world's population participates effectively in the modern economy, while a majority live near or below the subsistence level. Inequality of nations becomes graver as the world becomes more and more integrated. In the last decade, many peoples have become poorer in absolute as well as relative terms. A lucky few catch up by systematic borrowing of technology; more tend to lag. Meanwhile, some four-fifths of babies are born into societies unable to prepare them to participate in the modern world.

The situation of the poorer nations is difficult in ways hard to remedy: partial and unequal modernization increases inequality and division of their societies; financial dependence negates economic growth; many of the educated and ambitious emigrate to lands of greater opportunity; local cultures shrivel as mass culture, both highbrow and lowbrow, becomes global; civism gives way to corruption, uncontrollable in the prevailing anomie. It may be concluded that bad government is the principal, or at least most apparent cause of the low levels of technology, productivity, and wealth of most of the world. But no remedy is at hand. The problem of keeping government honest and reasonably efficient is difficult in all modern states; under the strains of weakness, poverty, demoralization, and dependence, the state becomes the strongest instrument of injustice and impoverishment.

To achieve a civilized eden, humanity must learn to reconcile ever-growing destructive power of nations with their sovereign freedom to injure one another; to accommodate individual freedom to reproduce and consume within the limits of earth's resources; to reverse the prevalent tendency of technological progress to deepen inequalities of wealth and power; in short, to harmonize order and freedom much better than humans have yet been able to do.

There are no technical solutions for such needs. On the contrary, social demoralization strikes at the roots of science. Increasing numbers of people in the most modern countries, in the U.S. and Western Europe, are profoundly skeptical of the ultimate benefits of much more technology, and there is distrust of the deeper manipulation of nature—shown by the allergy to nuclear power and fears of gene-splicing experimentation as well as passionate defense of threatened environments.

Interest in science may also wear out as the frontiers of knowledge are pushed farther from ordinary comprehension and as basic discoveries become ever more recondite and fewer. Concepts become more and more difficult and synthetic, if not of problematic applicability, as professionals strive to prove their credentials by profundity, which may take precedence over truth. Would-be investigators have to cover ever more ground before they reach interesting territory; one day, it may no longer seem worth the labor. Many modern mathematical investigations are accessible to a few dozen, or at most, a few hundred specialists.[27] A large majority of scientists work only on practical application of theoretical knowledge. But if science grinds down, it will be difficult even to maintain the inherited store of knowledge.

Successive technologies are more capital-demanding and usable by a smaller proportion of the earth's peoples. For example, the production of energy has advanced from felling trees to mining coal to drilling miles into the earth for petroleum and to extremely expensive nuclear energy. If fusion power becomes feasible, generating plants will apparently have to be too big and costly for all except the very richest nations. Technology creates ever higher dependency on fragile systems. Agriculture would collapse in the advanced countries without chemicals and fuel. A failure of electricity would create a life-threatening crisis, especially in cold weather. We approach the point that a few computer crashes could cause havoc to the economy. Coming decades are likely to see more expenditure on adapting ourselves to increased crowding and shortages than on climbing new peaks of technological grandeur, and rapid growth of material welfare may be coming toward exhaustion. Already, technology is failing in that the improvement of productivity and means of communication and data-processing does not bring commensurate improvement in conditions of living. The machine outpaces the human; and much, perhaps most, of its potential is wasted.

Moreover, high civilization requires an ever larger base. A village can sustain a stone age culture; early civilization, as in Mesopotamia or Egypt, could be self-sufficient or nearly so in a few thousand square miles. Western culture grew up on the scale of a small continent, or about a million square miles. Nowadays, because of the concurrent needs for specialization, competition, and large resources for production, exchanges are necessary on a very great scale in order to sustain dynamism. No country alone is large and diverse enough to maintain an adequate flow of information and innovation or to produce all that modern industry needs. At some stage, the smallness of the earth itself may restrict the attainable level of civilization.

But the dynamism of modern civilization continually pushes to new frontiers, and we cannot imagine a highly geared society settling down to stability, that is, to a succession of generations without major improvement. Whether cultured humanity can resign itself to modest growth without becoming wholly paralyzed is doubtful; stasis would probably mean regression. From our present observation-point, there seem to be stronger reasons to doubt than to hope that the overall trend toward higher order in the universe will go much further. If technology were lost or development were negated, it would be a little like the universe turning its expansion around and closing irreversibly back on itself, losing the order attained during the expansive phase.

If this should occur, our species presumably will regress to a much more primitive but self-sustaining level, like that of nearly self-sufficient villagers of Africa or Asia. This would be a biological and human catastrophe, worse than the periodic crashes reducing the population of arctic hares or lynxes by nine-tenths. The course of civilization cannot be halted at the present level or reversed without disemboweling civilization and condemning most of the 5 billion members of our kind to penury and death.

## THE ENIGMA OF PROGRESS

Humanity does not deeply understand itself and its institutions, least of all the political and social modes that make or destroy the effectiveness of society. We are swamped by too much information but know too little of what we really need to know. The interrelationships of countless factors are inscrutable; the more we study social affairs, the more intractable they become.

Yet to insure the continued prosperity of our civilization we should know why some groups have functioned well—in terms of creativity and productivity, the ability to apply intelligence to social purposes—at certain times and places in the past. We particularly need to comprehend why in our day some countries seem much better prepared than others to generate and use knowledge.

The historical record shows broadly that loose collections of independent states have made possible more open societies, in which decisions were decentralized and intelligence was unleashed. But openness of the international order and political systems is only a necessary condition of the creative society, not a sufficient one. At best it is a partial answer that does not reveal why the few highly productive state systems of history arose when and where they did. There are countless puzzles; for example, if we suppose that the virtues of the Greeks were promoted by what one might call competitive geography (the many little islands and peninsulas trading and communicating with one another by sea), why did the brilliant culture of Mycenaean Greece collapse a few centuries before the rise of classical Greece? And why it has been so difficult to spread Western culture, with what we see as its virtues and material benefits, although the model has been available to the world for many generations? In the course of the last decade or so, the standard of living of many nations in the Near East, which were in the vanguard a thousand years ago, has stagnated or fallen.

Most of Africa has fared even worse, despite the intensity of international exchanges and many well-intentioned attempts to help.

Argentina up to 1930 was one of the most prosperous and progressive of nations. With magnificent fertile expanses and a fairly homogeneous population largely of European background, it grew rapidly from the 1860s in riches and the ornaments of culture, achieving strong leadership in Latin America. It also established a promising democratic system. But with the coming of the great depression (although partly for other reasons), Argentina fell to a military coup in 1930. Since then, through a series of military regimes, Peronist dictatorship, and semi-democratic interludes, Argentina has sunk from the front rank to middling status among nations, with humiliating chronic economic weakness and political disorder.

Intellectual-cultural-economic success likewise eludes understanding. The Germans, for example, have for many generations shown themselves as though specially gifted by nature in arts and sciences. Prior to the First World War, they led the world in modern industry, chemical and electrical; and by organization and technology they came near overcoming the heavy numerical superiority of their opponents in two world wars. Recently, West Germany, with only a quarter of the population of the United States, has become the world's largest exporter. Yet German society has been overladen with authoritarianism and was (until after WWII) hardly open and democratic in temper. German cultural-scientific excellence has been accompanied by political narrowness or political misfortune, as shown by the militarism that contributed to WWI and the descent in 1933 into a tyranny unexampled in the history of the West.

Japan, alone of the many nations assaulted by the Western cultural, economic, and political onslaught in the 19th century, reacted successfully. Under an authoritarian but not despotic government, the Japanese adopted Western modes so far as deemed useful and adapted themselves to the modernization that promised power. Their success story has been much studied, but it has been described rather than explained: the Japanese, with their competitive-cooperative, technologically-oriented society, study and work harder than almost any other people except the Koreans. No doubt they profit from the modern political institutions imposed on them after WWII; but their postwar economic upsurge is only a continuation, after recovery from the disaster of the war, of a curve rising for well over a century.

Why should the (South) Koreans be at least as hardworking as the Japanese? Their nation lacks the background usually associated with proud achievement: it was long abused by neighbors, suffered as a Japanese colony from 1905 until 1945, and has been a dependent client of the United States since then. Korea at least has a homogeneous population, but Singapore is very mixed, with a Chinese majority. Yet Singapore, without benefit of democracy, is the most modern and prosperous of hot-tropical nations. Its people are more than ten times as productive as those of Mainland China—as are those of another "economic miracle" country, Taiwan.

In the important case of the United States, Americans are not much troubled to seek explanations for their success. Conventional reasons are God-given land and mineral resources, the ideals of the American Revolution, the wisdom of the framers of the Constitution, and the virtues of the people. In this century, victories in two world wars with small losses gave this country a fortuitous advantage. The primacy of the United States may, however, be fading. For many years, its share in world production and exports has been slowly shrinking, and the century-old boast of having the world's highest standard of living is outworn, as several European countries and Japan have, with much inferior resources, achieved similar income levels. Yet no one can say whether the ways of the past are no longer good enough, or Americans have declined in virtue, or other peoples have somehow improved.

Understanding of the dynamics of civilization is not much advanced by trying to relate to simple factors or universal principles. The chemist can have the moral satisfaction of knowing that macromolecules interact according to the forces analyzed by quantum mechanics, even though this knowledge does not ordinarily help in practice. The biologist's theory of selection seems to account for a good deal of evolution, even if it is inadequate as a general explanation. But civilization is more obviously mysterious. Historians seldom attempt theoretical explanations, like tellers of thrilling tales who do not tarry to psychoanalyze their characters. The forces, conditions, and factors in the ordering of society are too complex for the mind, and the complexities grow at least as rapidly as the ability to deal with them.

It is fortunate that no one has developed a plausible oversimplification to explain how and why civilization arose, what has caused it to progress or regress, or what it must be if it is to advance in the future. In the yearning for thinkability, any fairly plausible explanation would cramp thinking. For example, Marx regarded himself as Darwin's counterpart in the study of social evolution, with a materialistic explanation relating everything to ownership of means of production; his economic class approach has narrowed the intellectual as well as political horizons of countless believers. Visions of history proposed by such as Vico, Hegel, and Toynbee are too obviously incapable of explaining anything of importance to do great harm.

The "atoms" of society are inscrutable humans, whose behavior is baffling both as individuals and in organized groups. For the understanding of civilization, one would need to know a great deal about material conditions, the biological nature of humans, and their built-in capacities and tendencies of behavior. But this would be only the beginning: the present rests on a long history, important turns of which must be considered more or less accidental, and on the idiosyncrasies of many individuals, whose ideas and actions can have wholly unforeseeable results. Who would have guessed that the second most widespread world religion would be born when a visionary Arab merchant fled from Mecca in 622 AD? Or that a great political fission was beginning when, early in 1917, an emigré agitator slipped back to Russia to take advantage of the fall of the Romanov dynasty?

Even where the unpredictable individual disappears in the statistical mass, our explanatory abilities are woefully feeble. For example, the sophisticated science of economics, nourished by hopes that it can help to make money, tells very little about economic development or about the management of a modern economy—the experts cannot agree whether interest rates should be raised or lowered. No sociologist can give us a clear idea of the reasons for crime or remedies for it, or explain the pathology of family relations, or the causes of scientific dedication and discovery. Political scientists, if their ideas are to be relevant to reality, must deal with essentially undefinable and unquantifiable concepts, such as power, authority, freedom, legitimacy, charisma, and ideology.

The complexity and unpredictability of the universe reach their climax in this endlessly convoluted web called civilization, in which every element is related to all others. A tremendous mass of information, both explicit as written and implicit as embodied in objects, behavior, and institutions, forms a reality far greater than the sum of individual components. This is an entirety with its own meaning, the climactic achievement in the order-building of the universe. Finding expression only in and through individual minds, it is ever more shared and generalized, as though a collective consciousness, a mind of humanity. It is the acme of abstract being and at the same time a powerful remaker of our material world, a realization of something of the metacosmos.

## SUPERCIVILIZATION?

Can this creation of countless intelligences, having risen so far above unself-conscious animality, come to understand and guide itself well enough to attain a much higher and firmer level? The clearest hints we have are negative.

A civilization only slightly more potent than ours would be able to traverse the distances between stars and the planetary systems doubtless circling around many of them, and also would presumably be able to adapt to a rather wide range of conditions. Presumably such a higher civilization would be driven by the impulse to self-preservation and expansion that is inherent in life as we know it; it would hence strive to project itself to distant planets and thereby assure its own survival. There has been ample time for a supercivilization to colonize the galaxy; and any organization that spread over astronomical distances ought to be practically immune to breakdown. But there is no evidence that extraterrestrial beings have visited this delightful planet.

Moreover, the search for signs of any broadcasting civilization has thus far been fruitless. If scientists listening with giant radiotelescopes to the noises of the universe, scanning millions of channels, catch no message-like signals in the next few years, one must wonder why there is only silence. There are thousands of stars within their range of reception; and we assume that a civilization on our level would communicate over great distances, using enough energy to be detectable at a distance of many light-years. Even if they were not intentionally broadcasting to the universe, it should be possible to watch soap operas at astronomical distances.

Electronic civilizations can hardly be numerous. Is nobody out there? There are several hundred billion stars in our galaxy, about half of which probably have planets.[28] If physical conditions were suitable in one out of a hundred solar systems, there would be at least 50 million potential earths, many of which must have been hospitable billions of years longer than ours. But it cannot be taken for granted that a suitable earth will produce intelligence, much less that intelligent life must bring forth a high civilization. The probability would seem, from the historical record, to be decidedly low. It is more likely, in view of the tendencies toward rigidity, sterility and self-frustration, that progress will halt far short of broadcasting capabilities. Multiplying all the probabilities, one may roughly reckon that one civilization capable of broadcasting may arise per million years in this galaxy.[29] If some of the guesses are wrong, there may be none at all. If so many planets fail to produce a long-lasting civilization capable of making itself widely known, the chances of our doing so seem minimal.

Yet scientists are optimistic. They mostly cheerfully assume that there are no insoluble problems in reconciling community and individual needs in our artificial setting and that the expansion of knowledge and capacities so exuberantly flourishing today will necessarily roll on. A researcher peering toward the vistas of the unknown hardly doubts that the flight of science can continue without end.

The stronger reason for believing in the limitless expansion of technology, however, is the faith that the creativity embodied in the cosmos cannot run out, that all the achievements of intelligence cannot go for nothing or fail to achieve a higher flowering. Just as life is capable of infinite increase, so civilization should logically grow to encompass more and more of creation, perhaps ultimately merging with the original intelligence that gave rise to the great show.

Is this impossible? Our civilization has at least the advantage that it is ever more self-aware and self-critical; and if people are educated to the needs of survival for their grandchildren, they may be willing to accept the necessary sacrifices to construct a viable social order. If major war can be avoided for a few decades, violence should become outmoded as a means of interaction among advanced nations. In the subsidence of violence, authoritarian ideology must fade away; it is conceivable that governments will become more responsible and consequently better able to cooperate without sacrifice of independence. The rivalry of sovereignties may then be expressed in terms of cultural, economic, scientific, and other achievement.

Human intelligence may be increasing in the lively modern environment; intelligence tests show substantial gains over the last generation in developed nations.[30] Genetically given capacities are already supplemented in many ways by synthetic aids. Ever more self-sufficient computers reach toward what is properly called artificial intelligence, becoming more versatile and capable both of cooperating among themselves and of designing new circuits and networks. They should be increasingly able to assist decisionmaking not only in technical but in social affairs. In the not distant future, the computer can conceivably come to the rescue of the rationality of society.

Students of artificial intelligence assume that fairly soon it will be possible to contrive devices with more or less human capacities and that not long thereafter robots can be made capable of reproducing themselves—at least, if supplied with parts. To expect them to produce and refine the silicon to make their own chips is another matter, but this too should not be impossible. Ultimately, it is assumed, the electronic progeny of humanity should be able to do everything their human progenitors can do and more. Protoplasmic intelligence would then be displaced and left to extinction like other superseded forms of life, or kept in "wilderness areas" set aside for the endangered species.

But humans may not choose to be replaced by robots or to surrender governing functions. Modest efforts to fiddle with the human genetic endowment alarm many people; the idea of frankenstein computers would be politically extremely difficult. People are not likely to want to construct electronic slaves capable of freedom and power. The only obvious reason to make fully self-directing and self-reproducing robots would be to resist severe physical conditions or to last indefinitely, as for interstellar travel. Humans will hardly prefer to set aside the reproductive urges built in by long evolution, passing their heritage to the brainchildren of the few experts.

It also would seem excessively difficult to construct robots capable of forming a viable self-regulated society. Far harder than building problem-solving intelligence is to give robots the right measure of something corresponding to emotions and purpose. Unless the post-human robots had a built-in drive to reproduce themselves, they would soon become extinct; and those with more such drive (we assume variation, as in natural selection) would increase their numbers indefinitely at the expense of the less reproductively-minded. The selective advantage of maximum individual reproduction would inevitably conflict with the necessary altruistic-cooperative instincts of a high order. If our robots were endowed with something like ambition to motivate them to activity, it would be hard to prevent them, like their creators, from seeking power for their own benefit or that of their offspring.

The replacement of humans by robots would be less likely than the more effective joining of human and artificial intelligence and the improvement of the human organism. It is more promising to build on an already well-functioning apparatus than to construct *de novo*; it should be possible to harmonize humans with a technological existence and to make them better components of the social superorganism without reducing their autonomy. When or if the bearers of intelligence begin redesigning their society and themselves to make their intelligence more productive, civilization will attain a fundamentally new stage.

Such an outcome would represent a continuation of the long tendency toward the prevalence of information over crude matter; and the values and achievements of the beings of the future would be more and more cultural-informational, less concrete-material. The expanding and deepening sphere of knowledge, far beyond the reach of any individual, would encompass an expanding part of the substratum of the universe. If humanity is to fulfill the order-building direction of the universe, it must make itself a bridge from the animal to a higher existence of a kind that we cannot even dimly imagine.

Some believe mankind immortal, except as the torch may be passed to superior descendents freed of the limitations of protoplasm.[31] Still—forever?—encased in the material universe, this lineage may reach ever farther into the realm of order in and behind the universe. In an unimaginably distant future the universe may run down and in effect die. But at current exponential rates of development, human intelligence will have achieved everything of which it is capable in a very few centuries, an infinitesimally short flash in the life of the universe. In terms of the age of the universe, or the solar system, our civilization is newborn; and the future is boundless.

Can intelligence possibly go forward indefinitely? Living substance has given rise to what some have called the "noosphere," the realm of thought, above the biosphere, the realm of life. This noosphere is abstract, unsteady, variable from place to place and from society to society, different in each mind. Can the noosphere become ever stronger and more encompassing?

The outlook is so clouded that one might consider the longtime prosperity of human civilization to be evidence of a very special role in the scheme of things inherent in the metacosmos.

4

# Mind

## The material/immaterial mind

Civilization is realized in and through a multitude of minds, each unique and partly independent, each with its own vision of reality and contradictions between an inner and an outer world, each forming a symbolic universe with its causation like that of the material universe but more complex, as ideas and feelings interact and are built on one another.

The mind contemplates not only the external universe but itself. Each level of organizedness has its own unanswerable questions of the relationship between substance and order: the uncertainties of quantum mechanics; the apparent impossibility of fully accounting for living nature in mechanistic terms; and the mind-body problem, the timeless preoccupation of metaphysics and the master question of philosophy.

A blow on the head can make a genius cease to function as a genius, and perhaps as an organism; but this does not prove that mind is fully explainable in material terms. The agglomeration of neurons called brain makes possible the mind but does not constitute the mind, just as proteins and organs make possible a living organism but do not constitute the organism. The functioning of the brain proves that a sufficiently complex and well-ordered set of switches—the synapses and other interactions between message-carrying neurons—can make possible intelligence and what we perceive as mind, with ideas, will, and self-awareness. But we are very far from being able to account for the mind in terms of physical components.

A somewhat simpler apparatus may have a mind of sorts; at least, owners of intelligent pets are inclined to credit them with mental processes not totally unlike their own. Apes display near-human emotions, and dogs seem to feel guilty for transgressions. It cannot even be assumed that protoplasmic information processors are inherently superior to those based on semiconductors of silicon or germanium. Already, only a few decades into the computer age, semiconductor devices do much that their programmers cannot foresee. The builders of "neural net" computers cannot fully explain how even a small aggregate of processors creates newly organized behavior of its own.[1]

If a computer were constructed with sufficient capacity to react in a manner similar to the brain, rivalling it in plasticity and the ability to observe and correct itself and having drives equivalent to human emotions, it would have unpredictable creativity; and one could hardly deny it something like conscious awareness.

There would doubtless be heated controversy as to how a high-keyed nonhuman intelligence should be understood. Some would prefer to think of it in terms of the interconnections, the switches and other components, although the intricacy of the whole would defy efforts to relate its responses in a deterministic manner to its structures. Probably most observers would insist that with sufficient knowledge it should become predictable in a wholly mechanical fashion. Others would assert that, at the higher level of interaction, new factors come into play. They would find it not only too complicated for complete analysis in practice but inherently not entirely analyzable, just as we may know the factors making the weather but can never predict it accurately. They would assert that one could most fruitfully try to understand the "mind" of our hypercomputer by taking it on its own terms as a self-guided entity with its analogs of will and awareness, much as we credit other humans with feelings like our own. Perhaps if we encountered such an apparatus on Jupiter, we would accept it as a sentient being, whereas if we constructed it, we would regard it as purely mechanical.

The equivalent debate in regard to the human hypercomputer has gone on as long as people have given systematic thought to such matters. On the one side, persons impressed with human capacities (and perhaps their own importance) have stressed the specialness of our kind and our superiority to the animal and material world. On the other side, some thinkers, repelled by superstitions and struck by the materiality or muckiness of life, have concluded that only the material is real. Despite lack of evidence or a coherent theory, a few among the ancient Greeks and Romans, led by Democritus, Epicurus, and Lucretius, argued that everything was composed of atoms, the variety of the world resulting from the ways they were hooked together. Such beliefs had the virtue of making the multitude of classic gods superfluous. However, they lacked any factual basis and were largely submerged by religiosity until the 17th century. Then the burgeoning of science, through the work of Newton, Galileo, and many others gave more of a basis for mechanistic explanation. The debate in biology, psychology, physics, and philosophy has continued ever since, often with much passion. Discoveries and successful insights into nature nourish hopes of simplification; but remaining mysteries, often brought into relief by increase of knowledge, always thwart these hopes. Because of the emotional stakes, the controversy is notably less balanced and logical than is appropriate to objective science.

Regarding the nature of mind, there are many theories or approaches. Some seek to find a compromise between the materialistic and idealistic, as with notions of mental and neural processes somewhat mysteriously proceeding in parallel. Basically, however, the debate has been between those who stress the ideal or spiritual essence and those who emphasize the material basis.

Persons of religious inclinations or interests are horrified by the idea that the mind is only a register of physical events and the soul an illusion, incapable of surviving the death of the brain. They find it much easier and certainly preferable to imagine a spiritual reality under a supervising God than a physical explanation for evidently non-physical events.

On the other side, desire to surmount the irrational leads to the counterfaith that nothing exists except material particles and their interactions, and that whatever may lie behind this material reality is unknowable and inconsequential for our lives. The conventional scientific view postulates that there came into existence, for reasons we cannot conceive, a large number of material particles endowed with a great deal of energy; and all that we can perceive results from the ways these particles interact.

From this viewpoint, understanding is to be sought in analysis. Nature reveals secrets when dissected; and scientists, having achieved great success by taking things apart, usually regard this as the only sound procedure. To deny the primacy and sufficiency of the material is to attack the foundations of a very successful, impressive, and enthralling enterprise. At best, holistic understanding, treating things as entireties, is more nebulous and uncertain, less communicable, and more a matter of subjective appreciation, as when one feels that one understands a person or grasps a theorem not in the sense of being able to prove anything but in feeling how the person or theorem fits into personal relations or the structure of mathematics.

The training and professional life of scientists are geared to simplification; they like clear-cut, answerable questions. They look to experimentation with tangible things, not to probably fruitless wrestling with ghostly entities, such as an indefinable mind on which it is impossible to experiment reliably. Although the existence of consciousness is as self-evident as the existence of matter,[2] behavioral psychologists would reduce the psyche to stimulus and response, making reflex arcs (which are not really understood) the key to the personality, much as natural selection is regarded as the key to evolution. Whatever cannot be quantified or at least registered objectively is to be treated as irrelevant. A mental representation is obviously central to understanding behavior, but the behaviorists would abolish it.[3] This is comparable to insisting that the homing instincts of the pigeon are to be investigated by dissecting the bird. Similarly, logical positivists, untroubled by the contradiction of the mind declaring its own nonexistence, simplify their world by insisting that only empirically verifiable facts (however one may verify them) are valid. Many social scientists strive resolutely to quantify their data and arrange numbers and symbols in equations. Their findings are sometimes impressive but often irrelevant, as in political affairs, where the most important realities are not exactly definable and not quantifiable.

The belief that mental processes are theoretically explicable in terms of electrochemical happenings in the brain, as most psychologists would prefer, is a valiant effort to exorcise the demons of the mind. It is reductionist, like the effort to understand compounds in terms of the component atoms and atoms in terms of nuclei and orbital electrons. Like reductionist thinking in general, it has the advantage of at least a claim to concreteness. It appeals deeply to the scientific mind, with its mission of dispelling the darkness of superstition and mysticism. In contrast, efforts to deal with mind as mind are usually sloppy and subjective.

But materialism is a dogma like a religious creed. It is only an assumption contrary to intuition and common experience, and it suffers the fallacy of "nothing but": if the mind rests on interactions of neurons, it is nothing but such interactions. That the material basis is necessary and limiting does not mean that it is the entirety. Insistence on the materiality of the mind clashes ironically with the conclusion of many physicists that quantum events become real as registered, whether by a measuring instrument or a conscious being.[4] Because perceptions consist of complex and unanalysable events mixed with quantum uncertainties, one might conclude that reality is unreal. In the words of a German physicist, "There is no distinction between substance called mind and matter."[5]

One may regard as primary either the perception of the material world or the material world that (as we are convinced) gives rise to the perceptions. But the antithesis of the solid universe and the insubstantial mind[6] becomes less antithetical when we realize that the universe in its essence appears to be decidedly ethereal, and many of the proud attributes of the mind are emulated or excelled by very material semiconductors ingeniously interconnected. That the material is mental and that mental is material are opposites which, deeply pursued, come together. If we could look down on the dichotomy from above, we would perceive the antithesis of matter and mind as only the extreme of the expanding duality of substance and form in the creative universe.

The mind's perception of its world is like the view of a sphere from inside, while others see the sphere—myself in my world—from outside. Both the subjective (internal and personal) idea of the surrounding walls encompassing my universe and the objective (external and impersonal) perception of the sphere as an object in the universe are valid. One may likewise either think in terms of the freedom of the will (as we make choices that seem to be fully ours), or in terms of causes for choices (as we seem to observe in others).

The mind, like the electron, is real yet vaporous. It is more process than structure, much as the electron is wave function as well as particle.[7] It is essentially indefinable, observed only by itself; and different terminology applies to mind and body. It is difficult to argue that the mind is dependent or independent because we cannot define the mind in the terms used to describe what it is to be independent of.[8] It is like the ever-varied and unpredictable refractions of a kaleidoscope. It resembles a turbulence set up by the interactions of countless neurons, like the swirling of an agitated liquid. It may be compared to a hologram,[9] insubstantial but impressive, matter-based but intangible. This comparison is the more appropriate because large areas of the brain, like the hologram, participate in single reactions.[10]

Understanding of the mind may be holistic, viewing it as an entity with its special character, or reductionist, regarding it as an aggregation of simpler entities interacting according to discoverable laws. The choice is a question of a preference for thoughts and feelings or circuits and neurons. The two sides are exemplified by the duality of psychiatry, which tries to cure disorders by talking and engaging the will of the patient (in neuroses) or mostly by drugs or even surgery (in psychoses). The pathology is treated ambivalently as a disturbance of the personality or as a problem of bodily functioning.

In the common view, the person has two fundamental and essential aspects,[11] body and spirit or soul, but this Cartesian dualism is not so much a description of fundamental reality as it is a means of conceptualizing the world. It is to some extent cultural and synthetic; in developed form it belongs to modern civilization.[12] The two aspects are not separable. An image on the retina becomes vision in the brain; a will or idea turns into an electrochemical message to muscles.[13] The physical affects the mental, and the mental affects the physical.

Much of the difficulty of the mind-body problem comes from seeing a sharp dichotomy, because the mind functions by drawing distinctions. Mind does not represent a distinct world or a separate sphere but the summit of a continuum of order-making and responsiveness, with rising levels of complexity and creativity. The whole of reality is not logically coherent, to be drawn into any simple frame; and there is no more reason to say that material particles are all of it than to say that nothing exists except as it enters the mind.

From the viewpoint of consistent materialism, only physical things can have physical effects, and mental states are an illusion. But convinced materialists do not treat humans as mere chemico-physical aggregates, nor even animals, so far as they are intelligent and sensitive. Mind, will, and awareness should be understood in terms of brain function so far as possible, but they are not to be totally comprehended in such terms: they may be as fundamental to the being of the cosmos as is the electron.

The mind-body or mind-matter problem is the more baffling because it requires that the mind examine itself. It is impossible for one apparatus to predict in detail the operation of another of similar complexity.[14] Even if one theoretically could trace all the circuits involved in a decision, this would require something much more complicated than the brain itself. Whatever would understand has to be more complex than the understood; and the more sophisticated the mind, the more impenetrable. Understanding also fails because the investigation or observation necessarily influences the observed: registering mental events changes them.

We may assume that any thought has its material correlate; but causation can be physical, chemical, physiological, or psychological. There is a hierarchy of events in the brain itself, from molecular processes to interactions of individual neurons to activities of columns or brain elements, over which stands the integrating mind.[15] The mind arises from well-ordered complexity, seemingly mixed with something like ordered turbulence or what is commonly called "chaos;" and one may reasonably see not only the brain determining the capabilities of the mind but the mind acting on the material world through its brain.

This implies that a mental process has material effects. The scientific mind naturally finds this conclusion repugnant, although the idea that information modifies substance is accepted in quantum mechanics.[16] This seems, however, to be the way the cosmos is built, perhaps the only way that a material universe can become self-observant. It may seem intuitively illogical or improbable, but ours is an improbable cosmos; many necessary facts, such as the exactly suitable ratio of gravitational to nuclear forces, are utterly improbable. Unlike the alleged phenomenon of psychokinesis, the autonomy of the mind and its ability not only to direct muscles to action but to alter the functioning and conditions of organs of the body do not violate physical law. Mental causation is diffuse, entirely untraceable, and not fully predictable even theoretically; but its unpredictability is hardly more bewildering than that of the electron or the stock market. Mental processes are neither deterministic nor random but purposive.[17] Like the electron, the mind is partly self-determinant; and it has to be understood in its own terms as interaction of the metacosmos with the material world.

## MIND IN BRAIN

To fully understand the brain, in its complexity and multiplicity of elements, is a very faraway hope even for the most ardent materialist. It is a monumental achievement of the genes to put together an efficient network composed of the roughly 10 billion neurons of the cerebral cortex and more than ten times that many in the whole brain.[18] There would seem to be far more than needed for any practical purposes; crows manage rather smartly with only about 1/1000 as many. About 20% of the neurons deal with language-speech,[19] which is, of course, central for the mind. The bulk of the mass of the brain, however, consists of glial cells, the functions of which are obscure except for nourishing the neurons. A neuron interconnects with a hundred or a thousand others; there are at least 100 trillion switches or synapses. Neurons of the cortex are formed into columns, with dimensions of a millimeter or less, which seem to be information-processing units.[20] The chemistry of the brain is formidably complicated; over 50 different chemical neurotransmitters are used.[21] The fact that more than a score of drugs affect emotions, perception, coordination, fantasy, hallucinations, judgment, and so forth[22] suggests something of the complexity of the organ.

The brain is made up of numerous rather distinct parts—cerebral hemispheres, stem-brain, thalamus, hypothalamus, hippocampus, cerebellum, cortex, and others, all with different inputs, although they are mostly composed of the same kind of cells. It is realistic to speak of a person having "brains"; at least, there are fairly distinct subsystems, including movement, sensory perception, memory, decision making and planning, emotions and drives, and personal identity.[23] Some of the modules are in touch with language, others not; some are conscious, others unconscious.[24] Intelligence may consist of multiple faculties,[25] giving the human the most varied potential excellences, from the gifts of a pianist and gymnast to those of a statesman and scientist. On the other hand, there is also a general all-purpose ability to handle information, and people who are gifted in one area are likely to be competent in others.

This formidable organ, often called "the most complicated thing in the universe," is favored by nature. Comprising only about 3% of the mass of the body, it receives 30–40% of total blood flow, and consumes about 20% of the energy expended by the resting body. It is specially protected by the blood-brain barrier from chemical disturbances to which the rest of the body is subject. It is also difficult to experiment with. Knowledge of functions of its parts comes mostly from analogies with animal brains and as a by-product of surgery and accidental injuries. Much has been learned about it, but nothing that sheds much light on the makeup of mind. It is remarkable how little is known. For example, the location or organization of something so fundamental as memory is obscure, although large numbers of neurons over considerable areas seem to be involved.

What is known of the geography of the brain gives intriguing hints. If the big bundle of fibers connecting the two hemispheres (corpus callosum) is cut to check epileptic disturbances, the patient seems to have two minds,[26] although the functions of right and left hemispheres are somewhat different and only the one with the speech center (usually the left) can express itself verbally. Nerve impulses being essentially electric, stimulation of parts of the brain by minute currents gives some idea of their functions, and it has been possible to map motor and sensory areas in the cortex in some detail. An electrode in one part of the thalamus causes intense pleasure; exciting entirely similar cells nearby causes pain or repugnance. Stimulation of certain areas gives the subject vivid memories or flashbacks, but these are quite unpredictable and usually unrepeatable. The patient perceives them as extraneous and continues to be aware of present reality.[27] A limb may be made to flex automatically; but the patient feels that the movement is caused by the surgeon, not by the patient's volition. It has not proved possible to use an electrode to activate the mind itself;[28] and its location, if it has any clear-cut location, is unknown. Perhaps it moves about in the brain as attention shifts; or there may be, strictly speaking, no "mind" but a bundle of coordinating activities.[29]

The mind/brain not only receives information but generates activity, both as a whole and in its parts. Nerve cells do not merely respond to stimuli, like computer gates, but are self-excitative, spontaneously active in a coordinated fashion.[30] Their electrical potential is registered as brain waves, flashing many times per second in different patterns. When they were discovered a few decades ago, it was hoped that they would unlock many secrets; but they have thus far provided little information beyond generalities of the condition of the brain. They register attention, surprise, and other mental states; and they slow in sleep.[31] Their cessation is called brain-death.

The mind seems to be a sort of fusion of partly independent functions, at the same time consisting of emotions, sensations, and thoughts and dealing with them. It is oddly unitary, although the brain carries on many functions and we have an indefinite number of often conflicting wills, weaker or stronger, passing or permanent. The mind has a generalized ability to sort out and respond to diverse needs or calls for attention and is also a central command center to farm out tasks to different suborgans.[32] It is subject to emotions; yet it can, especially through training, largely master emotions; actors are said really to undergo the feelings appropriate to their role.[33] One struggles with the speech mechanism to find a word, or with the memory to recall an incident, or with an impulse to restrain it. The mind turns around and tells itself to be more alert, or to try harder, or to go to sleep; and it may or may not obey. What are called "voluntary" actions usually conform to its directions, but all too often we do things our higher or more cultivated selves do not want to do, and perhaps say, "The devil made me do it!"

## DOMINANT MIND

The mind is not merely a reflection of bodily events. If the chemistry of the brain influences the mind, mental states also have much influence on brain chemistry.[34] That is, there is mutuality and downward causation in the mind-brain relation. The mind's being an unmoved mover does not imply absence of mental causation,[35] but it means that free will is a reality. For us personally, it certainly is. The feeling of freedom of choice is as real as the feeling of desire. But if we see a dog hesitate between obeying the master's call and chasing a squirrel, we may take it that its integrative apparatus is momentarily balanced between conflicting urges, one learned and the other instinctive; and we are not troubled by any mystery in principle.

In the more complex behavior of humans, we have no hope of fully understanding outcomes either as results of external conditions or as caused by unanalyzable processes supposedly going on inside their skulls. But we can choose to believe either that a person's behavior is in some obscure way determined, or that it is an unpredictable, independent manifestation of an autonomous self like our own. A molecule is made up of atoms, which determine how the molecule can be put together; yet the shape of the molecule determines how the atoms behave and interact with other atoms and molecules. An organism is made up of various parts, on which it depends, but which it can to a degree control. Workings of the brain are to the mind as organs to the body, interdependent and mutually determinant.

A rational view of the world must harmonize with the intuitive reality of free will. That people are responsible for their actions is taken for granted as a practical matter, and it is difficult to imagine a legal system functioning without it. Behavior, after all, and what we see as the results of a free will are the chief meaning of human life; and minds, in their freedom to be good or bad or to go in any of a thousand ways, are much more different than bodies.

Difficult as free will may be as a logical concept, the alternative of determinism is even less manageable; it requires an exercise of free will to deny free will. It is somewhat paradoxical to argue that the mind is wholly determined by physical structures and events, hence that physical structures prove their own physicalness.[36] A striking fact of the mind is its urge to uphold its own identity; people strongly desire to assert themselves as independent non-mechanical individuals.[37] The strong personality demands to be captain of its fate, like the condemned soldier who would have the last freedom of giving the execution squad the command to fire.

Free will evidently presents a parallel to quantum indeterminacy. The latter cannot, however, be taken to explain the former. Rather, quantum irregularity would represent disorder,[38] unless one envisions a mind somehow able to choose among admissible quantum outcomes to cause nervous impulses, an improbable relation of the nonphysical to the physical. It would appear more reasonable to suppose that what we sense as free will is analogous on a higher level to the indeterminacy of the particle on a quantum level, an indeterminacy inherent in the creative complexity of the central nervous system and the nature of our existence, akin to the unpredictability that pervades the cosmos.[39] But the indeterminacy of the electron is scientifically respectable (although it was much resisted)[40] because it can be defined by equations and is confirmed by experiment. There is no hope of an equation quantifying mental probabilities, and no experiment can prove free will. Yet the autonomy of the mind may be as fundamental to the nature of things as the unpredictabilities of quantum mechanics.

That the autonomy of the mind is not illusory is suggested by its power, conscious and unconscious, over bodily functions. The mind is not only subject to ills of the body but may cause them. Some malevolence of the lower levels of the psyche can bring on a great variety of psychosomatic ailments, from sterility to colitis, perhaps even cancer and diabetes. Skin pathologies are especially likely to have a mental origin.[41] Emotional maladjustment damages the immune system.[42]

Most illnesses are to some degree psychosomatic in that resistance is much influenced by mental condition.[43] Grief causes susceptibility to infection, and an infant whose biological needs are amply filled will waste away if deprived of affection. On the other hand, those whose minds are stimulated probably grow better brains; at least, rats raised in an interestingly enriched environment develop a larger cortex, with more abundant connections, than their unstimulated siblings.[44] Primitives who believe in the sorcerer's magic may be killed by it; and some persons are said to have brought on their own death by act of will.[45] On the other hand, medicines are more effective if taken with faith, and placebos cure many ills. Pain may be quite repressed by something so apparently irrelevant as needles stuck into various parts of the body distant from both brain and the source of pain, permitting major surgery without anesthesia. Warts are regularly made to disappear by suggestion. Wise physicians treat their patients not only as bodies to be medicated but as persons to be encouraged. Those who do not care to live are much more likely to die.

The mind, in hysteria or otherwise subconsciously, sometimes seeks a way out of emotional problems by paralysis, blindness, or other psychogenic ailment. Similarly, words spoken by a hypnotist may give one unsuspected strength, make white look black, recover buried memories or erase fresh ones, cause the subject to do odd things without knowing why, or raise blisters. Hypnosis can bring about enlargement of breasts. Hypnotic suggestion of a fatty meal causes secretion of the appropriate enzyme (lipase);[46] the idea apparently serves as stimulus for a conditioned reflex. Somewhat similarly, the immunological system (at least in rats) may be suppressed by a harmless taste associated with an immunosuppressant drug.[47]

Through hypnosis, self-hypnosis, or training with reinforcement, many things not normally subject to the will can be brought under its control. People can learn to reduce blood pressure and dilate arteries, to alter intestinal peristalsis, or to change temperature of body parts.[48] An expert can reduce oxygen consumption for a considerable time to less than half normal.[49] Devotees of yoga perform a host of such feats, which modern science has hardly considered worth investigating. Subjects can learn to modify brain waves,[50] whereas one might assume that the brain activity indicated by the brain waves would control the mind.

To attribute such effects to the power of suggestion is merely to give a name to the mystery, like diagnosing stomach trouble as gastritis. It is likewise unhelpful to speak of conditioned reflexes without a clear explanation of the signal transferred to the innate reflex. Unhappily, science knows little or nothing of the hows and whys of hypnosis, faith healing, acupuncture anesthesia, psychogenic pathologies, sexual deviations, hallucinations, and idiot savants—not to speak of consciousness, attention, how stimulation of certain regions becomes a sensation of pleasure, and even such mundane matters as why we sleep and dream, not only sleeping but dreaming being necessary for sanity. The very foundationstones of intelligence, learning and memory, although relatively accessible to investigation, remain mysterious. Neurosis is sufficiently enigmatic that there are about a dozen schools of therapy. Despite an immense amount of research, insanity is unaccountable. A vast array of mental phenomena, including not only some of those mentioned but mystic and religious experiences, personality changes, sundry trances, and many exceptional but not clearly psychopathic manifestations make no sense at all in terms of brain structures evolved by selection to serve survival. "Near-death experiences," in which persons suffering cardiac arrest recount details of the resuscitation as though having watched from above, are not easily dismissed.[51]

It is also suggestive of a high integrative capacity that the map of the cortex is not fixed. For example, if a finger is amputated, inputs from other fingers move into the vacated region; and practice can enlarge the areas given to specific functions.[52] The brain has a remarkable ability to do without large parts of itself; and one part can often take over functions of another, at the same time giving up more or less of its previous faculties.[53] Functions of damaged specialized areas may, within limits and especially in children, be reassigned to other parts[54]—a fact that complicates the idea of the correspondence of neurons with thoughts. Severe injuries destroying masses of brain tissue may have little effect on the personality. If the usually dominant left hemisphere is badly damaged, the right hemisphere may in time take over much of its functions, including the power of speech, if patterns have not become too hardened and especially if a strong will, something nonmaterial, is at work.

Psychoanalysis, considering not structure but function, treats the mind as something like a little society, with an administrator (the ego), pressed from below by the elemental drives of the unconscious (the id) and checked from above by the socially implanted conscience (the superego). This implies a sort of invisible superior coordinating referee, deciding what is to be admitted from the unconscious into dreams or to the conscious mind and overseeing the contest of id, ego, and superego.

However the driving and directing forces may be named or delineated, the mind guiding the body is the acme of downward causation, in contradiction to reductionism. Dependent as it is on brain and body, the mind is like the general of an army. He needs his staff and troops to keep him informed and to carry out his orders, and he can act only within a certain framework; but he oversees, asks for reports, coordinates, and commands. At least, he tries to command; but he, like the mind, does not always have his way: contrary to his orders and exhortations, his soldiers may take fright and retreat when he would advance.

## CONSCIOUSNESS

Just as there is no sharp cleavage between the inanimate and the living, or between the mechanical and the intelligent, there is no abrupt beginning of awareness on the evolutionary ladder.[55] It rises, we assume, by degrees from lizards to apes to poets. There is a continuum of degree of complexity from the simplest nerve impulse and response through reflexes of rising complexity and dimly conscious behavior to the highest creative reasoning. All animals with a central nervous system are capable of learning, usually quickly if within their normal environment.[56] After learning by observation, the next step is to save experiences and use them to reason. Modifiable, creative, purposeful behavior would seem to require a sort of interior model of the world, that is, a rudimentary mind. Perhaps consciousness arises from vicarious trial-and-error, like computer simulations, enabling animals to make choices without risk or exertion and leading to symbolic exploration of reality.[57] Some would have consciousness inherent in the substance of the cosmos. This is poetic, but the fact of self-observation is the key mystery of the universe.

The boundaries of the self-viewing self are not clear-cut. The self-conscious mind can negate instinctive reactions, but there are shadowy corners where the light of reasoning awareness variably penetrates but does not prevail. Beneath lies a subconscious, like the lava under a volcano, usually quiet, occasionally bubbling up, sometimes exploding.[58] The consciousness admits sensations, ideas, and memories that force themselves on it; and it has great difficulty in excluding undesired intruders, telling itself not to think about unpleasant matters. On the other hand, it reaches out to attend to particular senses, to pull up memories, or to focus on problems. It is not a single thing but more like a spotlight moving across a varied scene. It is a protean process, held together by short-term memory, not an organ. From the outside, the mind is only inferred by analogy; from the inside also it can only partially perceive itself. The inner self-view is frequently deceptive, as when we realize that our motives were not what we believed.

Consciousness or mind deals with uncertainty of response; when one has decided, the matter usually drops out. Conscious attention is needed for learning; learned performance may be automatic. If one acts more or less mechanically, without being aware of one's actions, as in a trance, one has no particular feelings and can make no new decisions or plans in this blank state.[59] It is also possible to realize only later that one heard or saw something, or perhaps acted on a subliminal perception.[60] It is suggestive that the nondominant hemisphere in a personwhose brain has been surgically divided can make quite adequate responses to instructions without any evidence of self-awareness.[61]

I have a body but am a mind, while at the same time having one, that is, possessing myself. Although mind is antithetical to body, the concept of the separation of self and external reality is not innate. It is less obvious to primitive peoples; and children learn to distinguish self and non-self[62] while still endowing things with a will of their own—as we all continue to do in some degree. Language is probably necessary for clear self-perception. Some consider self-awareness to be strictly a cultural invention.[63]

The "I" is both actor and acted on: "I fooled myself." The observing self retreats: I am emotions, but I have emotions; the I is what is aware of the emotions, just as it is aware of a headache, or the imagery in its imagination, or the sky above, even as the I is moved by feelings. One says of the mind that it lacks musical ability or has a bad memory for faces. Some parts of the body are closer to the mind than others; one can envision living without a hand or a foot, while the stomach is more intimately part of the self. In a sense everything outside the thinking brain belongs to the outside world. If the brain were transplanted, one would assuredly become a new person. But what if part of the brain were substituted? For the meditator, anything outside the tight focus of concentration is an intrusion, and everything but the part of the brain engaged in meditation is an object.

On the other hand, the outer boundary of the self is indistinct. Things become extensions of the personality—clothes, tools, favorite ornaments, and so forth, like the sheath of the bagworm or the hermit crab's shell.[64] In a sense, one's territory, or everything over which one exerts some control, is part of the personality, the sphere in which the mind operates. The mind in a sense merges into the corpus of human culture.

Consciousness, the irreducible ultimate self, is more than a monitoring of brain activity; it not only feels but somewhat mysteriously knows that it is feeling.[65] Yet it is not possible really to observe one's own mind; as soon as one thinks about it, the contents of the mind are changed. The idea "he thinks" is probably antecedent to the conclusion by analogy "I think," from which it follows that I can think about thinking. By extension, we are aware that we are aware, and are aware that we are aware that we are aware, and so on indefinitely. Yet the idea would hardly occur to most people if it were not formulated for them; and in fact, after a few degrees of awareness, the mind loses track and is only repeating a formula. In any event, in ordinary functioning the brain does very little thinking about its own thoughts or looking into its inscrutable mirror.

The level of consciousness depends not merely on the wiring or innate capacity of the brain but on the information stored and manipulated, its quality and amplitude. This is the software, on the basis of which the brain, like computer hardware, operates. Experiences are refracted by the lenses of culture, the past of the personality, and its interactions with the corpus of human knowledge, as they enter the brain; and responses are similarly shaped. The mind has grown up concurrently with civilization. We may congratulate ourselves on a fuller consciousness or ampler mind than our paleolithic ancestors—the richness of whose mental life, however, we should be loath to underestimate.

The exalted capacity of the mind is not inborn but has to be built up, as it absorbs and digests material from the tremendous mass of more or less readily accessible information. Intelligence is not cognizable apart from learning; no intelligence test can be made independent of education and experience. The shape of consciousness is doubtless partly inborn; but it also is to a high degree acquired, like language. Something of a basic grammar seems to be wired into the brain,[66] but it is only a potentiality waiting to be filled out in accordance with culture and learning—and if one does not learn to speak in childhood, the capacity atrophies. Possibly the idea of selfhood would be unachievable if one grew up without it.

The mind requires external inputs or stimulation not only to develop but to function at all. Without external messages, as in experiments in sensory deprivation, it becomes disjointed and soon begins spinning aimlessly, ceases to be able to think about anything at all, and goes wild. The effectiveness of the mind depends on the quality and quantity of inputs, to the point where it is overloaded by more information than it can handle. Solitary man is not only helpless but nearly thoughtless. Even a mature mind, if isolated for long, loses the capacity for consequential thought. The hermit produces profundities only by reworking information carried away from civilization. The very concept of personality is artificial, that is, human-fashioned; even feelings of loneliness, withdrawal, or revolt are formed with the mental tools given by our culture. Individualism and individual genius are inventions of society. Only on the shoulders of a thousand generations and in communication with others can one become a rational giant.

## OBJECTIVITY AND TRUTH

Closely related to the idea of self and non-self is the antithesis of the subjective and the objective, between what is in the mind and what lies outside. Things ranging from sensations of pain or pleasure, sorrow or joy, to perception of beauty, seem totally subjective, that is, personal and only indirectly to be shared with others. Such intangible things contrast with what we have learned to regard as the real world, which we assume others to perceive approximately as we do. Thus we have the sensation of red, which is totally different from the sensation of green, and we assume that others share this sensation. But no words can convey how they differ, and it is hard to imagine how grass and raspberries look to the color-blind.

There is no boundary, however, but a continuous gradation from vague awareness to concrete sensing[67], from the most subjective feeling to seemingly wholly objective perception, from "I enjoy beauty" or "I like beautiful roses," to "This rose is red." General self-awareness goes into observing one's actions, such as a tennis stroke, which is not very far from watching another's forehand. The feeling of self, like the data of the senses, merges into memories, which are residues of perceptions. Melancholy melts into malaise; and headache is not wholly different from an aching back, or a sore joint, or the pain of a bee sting. But the bee sting, reported by pain receptors in the skin, is hardly less real than the visual image of the bee, although others share the sight and the hurt is only mine.

We consider whatever is clear, coherent, concrete, and well-defined to be objective. It may be difficult to distinguish a dream from reality; but the latter is consistent with the universe in general, whereas the former is detached and unsupported. The desk is convincingly real because one not only sees it but can lean on it, and it occupies a regular place in the web of experience. We do not need the confirmation of others to tell us that the breakfast cereal is real; it fits the pattern of known existence. But is a melody a personal experience or an external object?

A conventional criterion is that the objective is perceivable by others, while the subjective is private. This is not wholly satisfactory, however. If one sees a pink elephant in the parlor, he accepts (or should accept) that it is a personal hallucination; if others swear they also see it, then maybe there really is a pink elephant out there. One might be skeptical, however, unless it appeared in anatomical detail to be an elephant and acted like an elephant, crushing a sofa it sat on. There are collective illusions. Numerous reports of flying saucers, for example, have not been accepted as proof of extraterrestrial visitations. Should one say that the space-filling "ether" of prerelativisitic physics was objective because it was widely believed? Did general conviction make witches real in the sixteenth century? On the other hand, people may share empathetically quite subjective feelings; the sufferings of a loved one may be nearly as real as one's own.

Is the objective more real than the subjective? Why not believe what is nearest and most evident, the intimately sensed truth of the mind? But does inner certainty mean deep grasp of truth, supernal insight, wishful thinking, illusion, or intellectual vacuity? One may experience real-seeming sensations from an amputated limb. There is no yardstick to measure subjective truth except perhaps the equally suspect ideas of someone else. If perception of external facts can be distorted by suggestion, as in hypnosis, it is far easier to confuse the inner vision. The mind that stares long and hard down its hall of mirrors will usually see what it most yearns for. The subjective is vague, if not unintelligible; the objective might be defined as that which can be carefully studied and communicated.

Everything in the mind has some reference to the objective world; the wildest dreams are composed of ideas and images of reality, scrambled and distorted. On the other hand, the clearest perceptions are only partial and distorted reflections of reality, modified by learning and experience as well as the organs of perception.[68] Much happens to the picture on the long journey from the retina through at least a dozen relays[69] to the optic nerve and into several layers of the brain, the visual cortex, and the integration we call the mind. As in Relativity and quantum mechanics, the observer is crucial to the observation.

Knowledge runs a gamut from the near and definite to the broad, abstract, and subjective, from awareness of self to the deepest generalizations, as language and symbolism form a shaky bridge between the mind and its world. There is no clear-cut separation between the sensation of thirst and the awareness of refreshing liquid. One sees the rain, recalls previous showers, and associates it with weather patterns. One generalizes about water, both from personal experience and from learning about its properties. More remote is the knowledge that water is a compound of two gases, hydrogen and oxygen, a fact that we take on faith, without decomposing water electrically, because of the general reliability of chemistry. Still more abstract is the idea that the electron shells of atoms of these gases link in such fashion as to form a compound with the properties of water. The workings of the orbiting electrons in turn are guided by the abstruse rules of quantum mechanics and regularities we call laws of nature.

The laws of nature, which are most precisely defined in physics, underlie properties of substances (such as reflection) and things we perceive (such as waves in the pool);[70] and they merge in generalizations of chemistry, of physiology, and biology, and whatever regularities it may be possible to deduce governing intelligent beings and their societies.

On the side of abstraction, the laws of nature are inseparable from the mathematics in which they are formulated and which has suggested much of their development. The logical consistency of hyperdimensional geometries suggests that such a basic aspect of our world as the number of spatial dimensions may be in a sense accidental (although planetary orbits could be stable only in three dimensions, assuming that the general character of laws of force and motion remained valid). A four-dimensional hypercube (tesseract) can be described as fully as a real cube of three dimensions, and one can learn to deal intuitively with four spatial dimensions.[71]

Physicists speculate that quantum relations may lie behind the very genesis of the material universe, that is, have validity independent of the particles governed by them. The extent to which abstruse pure mathematics (as of group theory) applies to the farther reaches of physics suggests that an abstract formulation of the mind is deeply related to physical reality.[72]

## MATHEMATICS

The crowning creation and most powerful tool of the mind is mathematics. This queen of sciences is the synthesis of everything provably true without specific reference to the material world, as broad generalities of nature merge and are expressed in abstract relationships. It is an intellectual castle reaching indefinitely into clouds of reasoning, a mental wonderland, perhaps the most impressive extension of the mind, and a glimpse of the metacosmos, infinite in its potentialities, like living nature.

The foundations, however, are on earth. It would be impossible to begin to think mathematically without experience of the consistencies of the real world, its continuity and conservation laws. Two plus three must always equal five because things are not arbitrarily created or destroyed, and the counting numbers are the basis of mathematics. Geometry can be treated abstractly or algebraically; but real (mathematically imperfect) points, lines, angles, and shapes compose the underpinning on which reasoning about the abstractions is built. The importance of geometry lies in its coincidence with reality.[73]

Not arbitrarily but with total rigor, mathematics rules the material universe. The world is her slave; if any phenomenon seemed to disobey the governing formula or equation, we would only blame our understanding. Knowing that the volume of an abstract cone is equal to one-third the product of the height times the area of the base, we would be dumbfounded to find a real cone that did not conform as closely as permitted by our instruments. The sum of interior angles of a triangle in our space can only be 180 degrees, and the area under a curve corresponds to the appropriate integral as exactly as we can measure it. In any polyhedron, no matter how contrived, the vertices plus the faces have to add up to two more than the number of edges. We even feel confident that mathematics provides a reliable guide to the early history of the universe.

Mathematics makes science possible; its abstract power is the key to mastery of the material world. Sometimes mathematics marches ahead of science and lies waiting to be utilized; sometimes it finds in scientific discoveries new fields to explore. It is the quintessence of order in or over the cosmos, governing the physical world yet extending farther and deeper. It is like the expanding universe in that ever increasing and potentially infinite complexity is built on a simple foundation.

Modern science began in the 17th century with Galileo's mathematical description of motion, and all science aspires to as fully mathematical, that is, exact formulation as possible. The laws of physics are mathematical; if the theoreticians produce the correct equations, the experimenters should fill in the blanks and discover new facts to be accounted for by new equations. Mathematics is the ladder from the simpler to the more complex, the means of translating knowledge of particles into more or less understanding of atoms and molecules, or of atmospheric data into weather patterns—an often excessively difficult ladder, but an expression of the interrelatedness of things. It is the two-way bridge from experiment to law and from law to application.

The relationship of mathematics to the material world is profound. Physicists try to deduce the deeper laws of matter less from experiment than from mathematics,[74] and they sometimes seem interested in equations as much for their own sake as for reference to reality.[75] Einstein, although not a mathematician, pursued a more mathematical than physical logic; and recent theoreticians are fascinated by symmetry, a truly mathematical concept. So far as the laws of nature transcend or are antecedent to this universe, as many believe, they are inseparable from the abstractions of mathematics.

Yet even in rather elementary propositions, mathematics shows results far from the intuitive. For example, $e^{i\pi} + 1 = 0$ (the number e, the base of natural logarithms, with the exponent *pi*, the ratio of the diameter to the circumference of a circle, times *i*, the imaginary square root of minus one, plus one equals quite simply zero). Absurd as it is to raise anything to an imaginary power, this links the five most important quantities of mathematical analysis in the most direct way conceivable. There cannot really be any square root of minus one in this universe, of course, but it is abundantly used in calculations about concrete things, such as electric currents; and *i* could be a physical reality of a higher dimensional existence.

Mathematics makes unexpected discoveries remote from material reality. For example, it has been learned that there are exactly 18 kinds of finite simple groups and 26 of what are called "finite sporadic groups." The largest group requires 54 digits to number its members, and it is derived from transformation groups in a space of 196,883 dimensions.[76] It is related to certain analytic functions and possibly to cosmological theory. Over 100 mathematicians contributed to the proof of the completeness of the description of the groups, a proof occupying 15,000 pages.[77] There are curious verities. For example, how many spheres can simultaneously touch a sphere of the same size in 24 dimensions? The answer confidently given is 196,560, a result learnable only with the assistance of a computer.[78]

One may see the mathematician as creating new things or as discovering truths awaiting to be found, like a seafarer coming onto lands hitherto unknown.[79] People obviously make mathematics; one cannot say that theorems are there until formalized. Before Pythagoras, there was no Pythagorean theorem (that the square on the hypotenuse of a right triangle is equal to the sum of the squares on the other two sides), but it is a fact of flat space—it was known as a fact of certain triangles long before the Greek geometers, and there are dozens of ways of proving it. The calculus was practically asking to be discovered when Newton and Leibniz came upon it simultaneously; and the calculus they formulated can have results no different from those of any other genius's calculus, aside from differences of symbolism and some mechanics.

The dichotomy between creation and discovery is unreal: the limitations or potentialities of reason are not to be separated from discoverable external truth. It has long been known in practice that four colors suffice to separate boundaries on a map, but it was only an isolated fact until an involved computer program, requiring trillions of computations, was devised to demonstrate the impossibility of drawing a map needing five colors. Theories of infinities (a la Cantor) are a credit to the artistic imagination; but if a result has verifiable consequences, one might equally say that the mathematician's work is discovery or construction.

Mathematicians are rightly proud of the rigor, logical perfection, and infallibility of their work. Ideally, they start from a few intuitively acceptable assumptions or axioms and then, step by simple step, build up a conceptual edifice, as impregnable as it is magnificent. The reality, however, is not quite so clear-cut. Proof itself is indefinable; it means mostly refined and educated intuitions expressed in carefully defined terms and following consistent procedures approved by mathematicians, who do not always agree on what constitutes a proof.

Discovery-creation in mathematics, as in science in general, comes before proof, often by guesswork. Intuition and conjecture lead the way;[80] proof follows to confirm and solidify ideas and to convince colleagues. Proofs of hundreds or thousands of pages may well be too complex to be really checked. Russell and Whitehead took 362 pages packed with dense symbolism to establish that one plus one equals two.[81] The logic could be reversed: the truth of the equation, which follows from the definitions of number and addition, supports the ratiocinations of the philosopher-mathematicians. Proofs by computer rest on faith in the machine and the program; the mind cannot follow step by step, as in Euclid's work.

Sometimes nature seems to be hiding its secrets. There are numerous conjectures that seem true but defy proof; for example, that any even number can be written as the sum of two prime numbers. There is, moreover, an inherent uncertainty in mathematics. It has been demonstrated (Goedel's Theorem) that any axiom system permits propositions that cannot be decided within the framework of that system. To make the refractory propositions decidable, one must enlarge the system, which produces new undecidable propositions.[82] The internal consistency of a mathematical system can be proved only by principles the consistency of which cannot be proved, and there is no guarantee of freedom from contradictions.[83] These theorems of undecidability and incompleteness have little practical importance—examples are not easy to contrive—but they are of enormous theoretical significance. They seem to be universally valid: no system is wholly consistent—that is, fully understandable—within itself, whether in mathematics, physics, biology, or the cosmos as a whole. It is a corollary that intuitive certainty is broader than formal proof.[84]

Frequently one engages in experimentation to conjecture mathematical outcomes, or asks a computer to make calculations in a structured way and display results. Combining creativity and discovery, the computer has opened up vistas of "experimental mathematics." When humans are unable to devise a complete demonstration of a proposition for all values, the computer may show that it is true up to billions or far beyond, therefore true enough for all practical purposes and most likely true to infinity. The experimental mathematician feeds in a few simple iterative rules and initial values and watches weird shapes appear on the screen, as though unpredictable but beautiful chaos is built into nature.[85] The graphic depiction shows results entirely beyond the human capacity to reason out—if the result is created rather than discovered, the creator is the computer. The structure of mathematics, forming a gigantic scaffolding of concepts, methods, and results, is like the wiring of a computer—indeed, much abstract mathematics goes into the design of highgrade computers.

Mathematics is unlike what is ordinarily considered objective insofar as it refers to no particular physical reality—perhaps to no reality at all, as in the abstract geometries with which mathematicians toy. But it is objective in that it is the same for an indefinite number of minds. In ways beyond our present understanding, complex mathematical relationships may have determined, in the first instant of creation, the kind of space we live in.

Mathematics lies near the heart of reality, and it is a paradigm of the building of order. One can put a mass of bricks together to build walls, porticoes, houses, and cathedrals reaching heavenward with flying buttresses rising to noble spires; likewise, with the elementary counting numbers as givens and applying a few simple operations step by step, the mind has constructed the edifices of mathematics. The discovery of mathematical truth is like the building of new order in the cosmos as, in infinitely complicated ways, atoms join, step by step, to make living beings and minds that can construct mathematics. Like intelligence, mathematics is at once a product of the growth of order and a powerful means of its increase. It is an interface of cosmos and metacosmos, where ideas stand over matter.[86]

## The expanding mind

Thinking is an exercise or an exertion, like running; and humans usually prefer to sit or amble. Most of the functioning of the brain is mundane, serving only for minor or temporary needs. But the mind is capable of extending upward into the universe of order, stretching to encompass more of reality, thereby increasing that reality; and some achieve their most intense pleasure in soaring into intellectual heavens.

As part of the self-realization of the cosmos, the mind reaches above itself to take into itself something of the reality in which it shares. It is a distillate of the universal order-in-being. So far as it is creative, it is godlike; and it carries on the joining of order and substance. Mind and matter are not clearly bounded discrete compartments but parts of a single whole.

The created becomes creator. With the growth of culture and conscious mind, the awesome inventiveness of life comes into a new dimension, and purpose blossoms in self-awareness and social being. The mind is not only individual but collective, participant in the immense web of information and artifacts called culture. It has expanded as the power of culture has been enlarged through the inventions of language, writing, printing, and computer-processing; and the mind is inseparable from its sphere of operation.

If something like intelligence-will-awareness gave rise to material existence, the mind, which is sentient, self-aware, and endowed with will, is akin to what is ultimately responsible for its existence. In a sense the mind produces a world, like the metacosmos producing the cosmos. It is understandable that mental causes can have physical results, that there should be an input of mind to the material world, which seems to be the work of something like mind writ very large.

It eases the mysteriousness of the mind if it is regarded as a bridge between the metacosmos and material substance. Perhaps mind is not so much created by the brain as discovered by it, somewhat as the structures of mathematics—which, like mind, are a means of finding order in reality—are discovered by the mathematician within himself. This would imply that the fundamentals of any mind—that of a human, or a whale, or an elephant, so far as these may have more or less of mind—must be much alike.

In studying the universe, one studies, by the anthropic principle, the conditions of conscious awareness. Intelligence may or may not be a deep purpose, but it makes sense to see consciousness as primordial, not simply as a sort of reflection of unconscious existence. If we can reasonably develop ideas about such questions as the purpose of the cosmos, it is because we incorporate some of the higher order of things; and we can hope that our sharing in it may extend beyond the material expression. This may have something to do with the religiosity that is a strong part of the human experience.

What the mind may achieve in the collectivity of culture and with the adjuncts it can provide itself, the synergy of protoplasm and semiconductors, has no imaginable limits. The order of the cosmos has been building for billions of years, and we can guess that its expansion will not be reversed, although we do not know that we are its best vehicle. It would seem difficult for the brain-based mind to transcend dimensions of space or time, but we cannot guess how far it may reach.

# 5

# The human way

## Purpose

The ape stood up, shuffled blinking into the glare of civilization, and looked around bewildered. It is part of the miracle of the cosmos that members of a species given to hunting and gathering should become planters, craftsmen, planners, and thinkers and gain enormous success thereby. In higher stages of this miracle, the onetime hunter-gatherer, who grew up in scattered bands of a few dozen, learned to live packed in cities of hundreds of thousands or millions. He even took a liking to this unnatural environment. Humans are not like quetzals, which die when caged, but like sparrows, which thrive wherever they can find food and shelter.

This self-trapped creature prospers but does not know its character or where it is going. In the roots of its being, it is an animal; in the sentient part, it is a mind or spirit, bound up in an immense culture; and the dual natures are perpetually at odds. We are miraculously endowed with the ability or burdened by the need to wonder, and it tells much about us that we are troubled by the unknown. We would be better off materially to ignore mysteries we were not destined to fathom. Like nonhuman creatures, we can live life without understanding it, just as we watch television without any idea what is going on inside the apparatus. Yet something in us is not satisfied simply to live but demands to know. We somehow have to ask the unanswerable, "Why are we here, after all?" or "What is it all about, so great and so puzzling?" We want to draw aside a corner of the curtain of appearances to glimpse the inner springs, or to find the key to the cosmic riddle, of which we are a part.

We may lay responsibility onto a God, whose intentions are mysterious. But we must occasionally ponder what our God may have had in mind, whether His or Her or Its intent was an experiment, a game, a joke, or (as we might prefer) an act of self-fulfillment, like an artist's masterwork. The first chapter of Genesis, it may be noted, gives no indication of God's intentions in creating stars, earth, and its creatures, except to state six times that He called His work "good," and finally "very good." Did He/She/It assign us a significant role in the immense spectacle, or are we just happenstance by-products? Are we only a small token of the creativity of the cosmos, unlikely to endure more than a tiny fraction of the span of the dinosaurs that once lumbered so majestically? Is our existence like the babbling of the brook, which signifies only the tumbling of the waters, or the breeze that bends the grass and passes on to nowhere?

A partial answer is to see ourselves simply as creatures of the earth. From the biological point of view, the good of life is life itself; its objective is to reproduce its kind. Some scientists would like to conjure away all philosophic questions by postulating that the sole purpose of the organism is to propagate genes. In this view, the egg is designed to make more (fertile) eggs; the chicken is only a means to this end. Otherwise expressed, the sole function of the limbic-hypothalamic system, which generates the emotions that color human existence, like other organs and the living individual as a whole, is to make possible the replication of the genes. As an engaged writer put it, "the organism lives not for itself but to reproduce the genes for which it is a temporary carrier. Every animal's role is to transmit, spread, and protect its own DNA."[1] Or, "the individual organism is only [the genes'] vehicle, part of an elaborate device to preserve and spread them with the least possible biological perturbation."[2]

This view is too narrow. Life cannot rationally be characterized as any single thing with any single direction. It is more reasonable to say that the purpose of the genes is to reproduce the organism than vice versa. The reality of a mouse or a whale is in the animal, not in the string of bases in the chromosomes of its reproductive cells, the only importance of which is to program the making of new individuals.

It is realistic to affirm that expansiveness is built into life by the evolutionary scheme. From bacteria to redwoods, from diatoms to whales and humans, life is geared to generate more life. If it were not, it could never have arisen; and if it loses that purpose it will cease to exist. However, the desideratum of reproduction does not mean mere copying, but propagation preferably with improvement. If simple self-perpetuation were the goal, the tree-shrews of our ancestry would have to be counted a failure, although the horseshoe crab is very successful. The egg ought to be happy (if eggs are capable of happiness) if by an error in self-replication of its chromosomes it hatches a superior chicken and a lineage of more effective egg producers. Life first of all perpetuates itself, secondarily improves its capacity for self-perpetuation. It creates the unpredictable with the purpose of making its future. The purpose of life is life, and perhaps the improvement of life; it creates mind, which serves life but also fulfills itself.

The biological imperative is valid for us as it is for fellow inhabitants of the earth. Since long before the days of Abraham, it has been held a great happiness—possibly the greatest happiness—to have at the end of one's life a large and prosperous progeny. Children are a token of immortality, a bridge to the future, as each generation receives from the past to give to those to come.

But procreation cannot be held to be the sole meaning of human existence. The physical capacity for producing children is far greater than our ability to give them a good life. In the richer countries, and to some extent in the poorer, reproduction is no longer limited by the scourges of disease and hunger that once balanced births with deaths; and unlimited multiplication would soon exhaust human and natural resources, bring technological progress to an end, and lead to a collapse of the human population. Just as the natural impulse to eat abundantly and store fat for times of shortage gives way to restraint of appetite for the sake of health, the urge to self-propagation has to be restricted. The biological impulse to have many offspring cannot weigh heavily in modern life.

The future belongs not to the individual lineage but to the group or the society. Life teaches not only competition and strife, the battle of each for its own survival and that of its progeny, but cooperation and mutual benefit, from the alga and fungus that join to make a lichen to gregarious herbivores and socializing primates.[3] A leading reason for sociality is security; the animals that congregate have a better chance to fend off predators, as a small flock of songbirds wards off a hawk. Social traits have two aspects from the evolutionary point of view: a group with cooperative abilities may be more successful in competition with other groups, and it may be advantageous for individuals to conform to group needs as the price of membership.[4]

The family is the foundation of animal as well as human society, and group ethics is an extension of family ethics.[5] In the huge families of social insects, which are the effective units of reproduction, the large majority of members are sterile. Individuals are of no importance; social bees, ants, and termites sacrifice themselves with total self-abandon in defense of the hive or colony. For humans, too, despite their individual biological lineages, the destiny of the community is nearly equivalent to the destiny of its members. The greater the sweep of technology, the more absolutely are individual fates fused. Probably our children, certainly our grandchildren, can prosper only in an environment of health and prosperity of humanity.

Animals are more than bodies; the spider is incomplete without its outspread web, and the well-crafted dam is a part of beaverhood. We likewise are fulfilled in our culture: a naked, speechless, toolless creature would be a poor human, and indeed could not survive without the charity of fuller humans. As life rises to higher levels of order and the individual is melded into a vast highly structured superorganism, its purposes also must be in harmony with its relation to the community and the species, if not to the cosmos. We do not escape the built-in drives given by the past and our basic nature, but it becomes our task to find new and better expressions of human destiny and fulfill whatever role we can find in the evolving order of things.

## PLEASURE AND PAIN

Most of our actions are determined by sensations that stimulate reactions to the world. But we are poorly served by our desires for pleasure, because we have leaped, in a geological instant, into a new, wholly different technological existence. It is impossible that we should be quite suited for this strange new existence.

The human character has many quirks and irrationalities inherited from the distant past. The sneer, a relic of a snarling threat, has outlived the menacing canines it was designed to exhibit. Hair sometimes stands up like the fur of an angry cat or dog trying to make itself more imposing. Fainting is a singularly inappropriate response to horrifying sights, and it is odd that grief should flood the eyes. Why do we yawn, and why is it contagious? What is the sense of a violent expulsion of air when one feels chilly? The body plan is ill adapted to upright stance, as sagging bellies and aching backs testify. We have a legion of foibles, such as bad tempers, jumpy nerves, silly egotism, purposeless pride, irrational moods of euphoria and depression, weaknesses for many harmful drugs we have found or invented, and a propensity to undo our prime gift of reason through neurosis or psychosis.

The farther civilization lifts us above inborn needs, the more doubts arise. Problems of choice grow with the complexity of the environment and the organism's controls, the higher level of culture and the greater application of intelligence. Our troubles are less intellectual than spiritual, less a question of how than of why. The mental capacity inherited from our paleolithic forebears is so elastic and so capable of building endlessly on accumulated information and ideas that it seems fully adequate for the technological age. But the inability to direct these phenomenal capacities becomes ever more frustrating and dangerous. Moral improvement has become more necessary than material progress.

Even in simple matters humans are disoriented. The wolf has a problem of getting enough to eat, not of choosing what to eat. We struggle, often in vain, with an excess of food. The sense of taste that would guide a primitive to a nourishing diet of fruit, meat, and grubs brings the apartment dweller obesity, heartburn, rotten teeth, clogged arteries, perhaps diabetes. People have to learn to choose vegetables and bran cereal, swallow vitamin pills, or perhaps deceive with artificial sweeteners the taste buds eager for ripe figs or berries.

We could not remain alive, however, if the pain-pleasure mechanism designed to orient learning and response patterns in higher animals did not usually serve adequately. The reflexes of pain and pleasure are an integral part of the process of learning (which psychologists call "operant conditioning") by which we get along in the world. So far as reactions are conscious and steered by desires, whatever is life-forwarding, favorable to the protection, self-maintenance, growth, or reproduction of the organism, should be desired and therefore pleasant. To the extent that fulfillment of vital ends is voluntary, learnable action, it should be rewarded by the sensation of pleasure; and ordinarily it is. Hurtful or annoying feelings, contrariwise, should be a whip to punish or a goad to drive, warnings of harm or demands for needed action. Our artificial society has not gone so far from its roots as to prevent this mechanism from guiding us through most of the business of living.

Life is propelled by a series of tensions, the release of which is felt as pleasant, such as an itch to be scratched or a hunger to be sated. But all pleasant activities are harmful in excess, and as the hunger is satisfied it is normally switched off until food is needed again. Unconscious, stereotyped activities, like heartbeat, intestinal peristalsis, glandular secretion, or breathing, need no conscious guidance and produce no sensation as long as they proceed properly. On the lower level the body has many means of automatic stabilization or homeostasis. For example, reflexes help maintain constant temperature by causing shivering or sweating and the closing or opening of peripheral capillaries according to the weather. On the higher level, discomfort, or anticipation of it, turns one to a coat or a fan, or, with more utilization of intelligence, to insulate the residence and pay the oil bill. The anticipation of pleasure (or of discomfort) becomes more important as a directive than the sensation itself.

The nearer life is to the level of subsistence, the more largely its pleasure is earthbound and natural; this is the virtue of poverty. The richer and more powerful people become, the more they look beyond necessities to a more varied and complex range of experience. Having made a usable pot, the potter looked for paints to make it pleasing to the eye. The hunter, having bagged his dinner, shoots a colorful bird to make a headdress. In our civilization, it is taken for granted that people devote much, sometimes most of their energy to activities with no clear relation to biological needs, from sports to tourism, stamp-collecting, philosophy, or goals such as prestige, physical and mental capacities and skills, and artistic creation.

The mind plays with symbols detached from reality, and the most varied abstractions and ideals command fierce allegiance and arouse intense joy or sorrow. Millions rejoice wildly because of the ability of a little group to knock a small ball around. Even without material benefit, power is addictive. Desires sublimated from material satisfactions become boundless and insatiable; this is part of the discomfort and enjoyment of being human. Often our essentially artificial pleasures are perverse. People sacrifice their health to drugs and risk their lives for thrills.

If pleasure is all too often deceptive and harmful to the organism, its opposite is usually truthful and salutary. Pains, aches, and discomforts inform about nearly all the ills of the flesh. Pain is efficient. Almost every tissue of the body has a warning system, the principal exception being the brain itself. In the woods, foul taste fairly effectively repels us from noxious eating. While civilization often distorts pleasures, modern medicine has luckily increased the utility of pain. The torment of a toothache, a heart attack, or a rupturing appendix is little help to a savage, but it hurries us to the dentist or hospital. The defect of pain is often in its useless excess, as when it may advance dangerous shock in an injured person. But when pain is a hindering distraction, as in excited combat, it may be suppressed.

Pleasure and pain are by no means simple poles, more of one being equivalent to less of the other. They show striking differences and are distinct psychological functions with separate but adjacent housing in the brain. The various pleasurable sensations are not so very unlike, despite the difference between a good concert, a satisfying meal, and a victorious game. There are quite diverse kinds of unpleasure, such as the pain of injury, vague aches, foul taste or odor, fatigue, hunger, frustration, and fear;[6] but all are designed to protect the sufferer.

Unpleasant signals carry higher voltage in the control network. Suffering is the outcry of harm or danger that must be attended to. No physical delight can mitigate scalding pain, and it is easier to inflict unbearable pain than to give moderate pleasure. In age or sickness, the capacity for pleasure is sooner lost than that for pain. The memory is more scarred by humiliations, fears, and sufferings than graced by recalled delights—it is more useful to remember dangers and mistakes than good fortune. No dulcet idyll is so sweet as a tragedy is bitter. Language has a richer stock of words for bad feelings than for good ones; and it is easier to compose a sad story than a joyous one, which is more likely to be silly or vapid. It is easy to paint hell in lurid colors, but the ineffable delights remain ineffable, and visions of celestial perfection are boring. In the Hindu philosophy, mere absence of desire and pain is the ultimate bliss.

Sweets soon become cloying, but more torture brings more suffering. Discomfort jogs and stirs, while comfort softens. We can do without pleasure, but we need pain. We like pleasure and detest pain; they are made to be liked and disliked respectively. Yet with experience and mature reflection we realize that sensuous pleasures add little to the balance of happiness, are easily submerged in tedium, and perennially demand novelty to regain allure. To really appreciate good eating we must be a bit hungry. At the same time, the easier pleasures seem to detract from the more subtle enjoyments that we feel to be higher or more worthy.

Hence, spirituality often looks beyond the values of the senses, even inverting them, calling pleasure sinful, illusory, trivial, wicked, or despicable. Pain is held to be virtuous when voluntarily accepted and stoically borne. The strong-willed have always considered it weak to be slaves to sensation, from the savage too proud to flinch when roughly gashed at initiation ceremonies and the fakir who sleeps on nails to the moderns who seek simplicity in primitive living. Not merely selfish or vicious or disorderly pleasures have come under religious or philosophic disapproval, but all robust delights—whatever seems to detract from what may be regarded as the higher values.

To adapt to this strange condition called civilization, humans have to learn to reeducate their inborn nature. Just as the pain-pleasure mechanism enables learning to replace instinct, reason takes over from genetically given drives and rises above immediate sensations. It is part of the price of power to have to master oneself. To prosper in a world for which we are ill-prepared, we are called upon to act in a sense unnaturally, or according to a higher nature, the human role in the continued creation.

## HAPPINESS

Civilization brings new woes along with new pleasures, and needs of the spirit multiply as needs of subsistence are filled. It is not provable that people are happier with the comforts, amusements, and conveniences of modern life. Perhaps the simpler existence of our great-grandfathers was less nerve-wracked and more satisfying. Possibly Papuans in villages little touched by modernity live in better harmony with their world, in closer kinship with their fellows, and with fewer stresses than average Americans. Do the richness, movement, and variety—above all the opportunities—of modern life compensate for the intranquillity and distortions? After the inhabitants of isolated and primitive Tristan da Cunha Island had been evacuated to modern Britain in 1963 because of a volcanic eruption, they nearly all opted to return to their former simplicity.

But simplicity has never been the lot of humans: the paleolithic hunter surely wondered why his neighbor's magic was stronger against the cave-bear, or who had cast a spell to sicken his children. The early sailor had to make precious offerings to bring good winds; Agammenon sacrificed his daughter. Civilized life is no great hardship, whatever the troubles of accommodation. Those who have become accustomed to modernity seem to prefer it; and even those who would retreat from it would take large chunks of it with them, from medicines to canned entertainment. Since the beginnings of humankind, better ways of making and doing things have made life easier, more varied, and materially more abundant, and have enabled hundreds to live where one lived before. The modern upsurge of technology has given to humanity, or to the fraction benefiting from it, not only doubts and worries but a new amplitude of life and experience and power over things. The artificial, after all, only corresponds to the capacities of our intellectual community.

If riches do not buy happiness, few are deterred by knowing this from wanting as much as possible, and most people spend the greater part of their energies earning money. Poverty is no happier than affluence, and the transition from the one to the other is very enjoyable. Wealth is like the civilization whose products it buys. At a price in anxiety and complexity, it facilitates doing and enjoying many things and opens up many avenues. The rich may sleep no better than the poor, but the poor long for the comforts of the rich even at a possible price of less sound sleep. Similarly, poor countries want modernization and industrialization, riches and power, although these may mean traffic jams, social ferment, and air pollution. Material possessions should promote the manifold expansion of life, like the organs and capacities achieved through evolutionary history. Like hands and eyes, possessions are a means to the to fulfillment of our being.

It is natural that humans should be materialistic; through acquisition and fabrication of material things the hominid rose above less gifted competitors. However, possessions often become an end in themselves rather than a means to the expansion and betterment of life. Material things are often wanted out of mere pride of ownership, vanity, or desire for status; it is human to esteem diamonds because they are scarce. But if we learn properly to use our enlarged capacities, they should carry us farther from the grubby existence of our distant forebears. We can overcome sicknesses, minimize suffering, leave behind fears of incomprehensible evils and malign spirits, broaden horizons, and raise the level of the mind—perhaps largely relieve ourselves of concern for material needs.

Happiness is not an abode but a way or a passing on a way, not a goal but a by-product. Like life and soul, it is process, not being. Never quite fulfilled, it has no halting but is unfixable as music. It can be upheld only by being built anew. We travel for the trip as well as for the arrival, and happiness comes along the road, not as a treasure-chest at the end.

This is inherent in the nature of life and in accordance with the ever growing complexity of the universe. For plant or animal, standing still is at best a time of waiting for more growth, activity, or increase. Humans hardly find happiness in status, comfort, or possession, but perhaps in their acquisition and increase. Things desired shed importance as they are attained, and having means wanting.

Happiness is elusive, but it is always possible to widen horizons and find adventure. It is the business of life to be vital and creative. We do not inquire whether Shakespeare was happy, but he must have deeply enjoyed the creation into which he put so much intellect—at a cost of immense labor, unless his genius was unlike other geniuses of whom we know more. Newton drew much pride and satisfaction from laying bare magnificent truths of the nature of things. But such passing feelings were important only as they contributed to the lasting achievement. The works live on, like the hard body of the coral.

Contentment brings spiritual languor,[7] and the lotus eaters create nothing. Humans need mountains to climb, challenges and difficulties to overcome. The worst enemy of happiness in the modern society is boredom and futility, tedium without purpose; and purpose is the balm of suffering. Life fulfills itself by projecting itself outward and forward, through people, work, and ideas, sharing in something greater than the self, with broader meaning in the world. The hunter goes for the chase, the builder enjoys construction, and the lover is driven by love, whether attaining happiness or courting vexations, always fulfilling a human destiny. Agreeable and disagreeable sensations help (or sometimes hinder) in dealing with reality, but they are only incidental to the life that has risen from the ooze to acute self-awareness.

The hope of happiness lies in one's best functioning as a human being. This is the more complicated and contradictory as civilization becomes more confusing and values more uncertain. We are animals with animal needs and drives, whose brains are first of all tools of biological survival. Bodily needs are primary, and health is almost prerequisite for achievement. The sound of body are usually sound of mind, and it is easier for the healthy to be happy. Selfish concern and individualism are natural and an essential part of the social complex. But we have a biological and social role as members of the species, as spouses and parents, whereby individuals are a link in the chain of being. The celibate existence sacrifices much of the meaning of life and the problems that add to its depth. Love and sex, idealization and carnality, typify the different aspects of being human.

We are social beings, whose lives have meaning in relation to other humans. We also share in something far above the animal world, with minds that defy the logic of animality. We have overlapping or frequently conflicting roles: as individuals, as family members, as participants in various groups, as components of humanity in many aspects and activities, and as atoms of the evolving cosmos. Roles grow farther apart as the structure of civilization becomes more complex, and success in one realm has its price in others. The great personage is rarely a hero to his valet, and few geniuses have had much success in their family relations. This contrariety is a large reason that we are less happy than our affluence would seem to merit.

Conflicting emotions are mingled with the longing for permanence, with a sense of the incompleteness of this existence and a hope that what has been so laboriously achieved should not be lost. But the tree of everlasting life was left behind in the Garden of Eden. No one can slay the dragon of time to clutch the jewel of immortality; and each is soon worn out, fit to be discarded. There is a season to plant, to cultivate, and to harvest; there is a time to grow, to love, and to depart. Life has continually to set out afresh, with new bodies and uncluttered minds.

Awareness of mortality lends poignancy to the experience of humans, unlike animals, which are blissfully unaware of their inevitable end. This may not be altogether true; dying apes are said to drag themselves away to seclusion, and elephants sometimes cover the dead with branches. But the knowledge or the fear of death compels the human spirit to look outside itself. The yearning for immortality enters into strivings for power, riches, and fame, the desire to create, perhaps to earn a place in the hearts of at least a few if not a ticket to heaven. When Horace sang, "I shall not entirely die," he referred not to his children but to his odes, which he hoped would help to educate his society.

Life is brief and its balance doubtful; victory and defeat lie close together, and life implies death. It is an unfulfilledness, a perennial writing and erasing, an eternal becoming, suspended between the memory of the past and the vision of the future. Few can be great or celebrated; triviality is the ordinary destiny, and we cannot expect the world we leave behind to have much thought of us. But human self-actualization, the expansion of the personality, harmonizes with the self-actualizing universe. This, the acme of the hierarchy of needs, is not to be strictly defined; but it means maximization of human potential.[8] It yields the joy of creativity, of giving wholehearted expression to something of the inner being.

The fates dispense happiness more generously than wealth and fame. Perhaps everyone, no matter how poor or deprived, has some share of happiness. But some seem to have larger portions of light. There is a contentment in merging one's being in something larger than the self, in contributing to values less frail and transient than the individual, and in revering the creative order in and over the cosmos.

Neolithic builders must have gloried in levering huge blocks into place to make temples quite as impressive to them as skyscrapers are to us. Modern folk likewise hope to do something of enduring value. All who can contribute to making the world a trifle better, strengthening the wings of intelligence or raising the freedom of the spirit, either directly or indirectly through others, should take satisfaction in taking part in the meaning and making of the cosmos, which includes our better selves.

## Values

Right and wrong as understood today are basically much as they have always been. Kindness, unselfishness, self-restraint, honesty, and decency are timeless values; and greed, hatred, anger, egotism, and sadism are evils, as they have always been, because the basic needs of society are much like they were a century or a thousand, perhaps ten thousand years ago. It is fair to say that the good person is one who helps make a good society; and the good society, resting on cooperation and traits making social existence agreeable and effective,[9] makes a broader and more promising future for its members.

Yet the old answers are no longer adequate; ideas of ethics need to be refined and improved in the new age even more than we need to reeducate our pleasures for radically altered conditions. This is most patent in affairs of sex and reproduction, which comprise much of what is ordinarily called morality. In a different way, patriotism requires qualification when a war between leading powers would be inexpressibly catastrophic. Martial virtues and love of motherland, for which so many millions have cheerfully died, cannot serve powerfully to inspire and integrate nations when quarrels between sovereignties are outworn.

The guidance of the past is also undermined by new perceptions of our universe and ourselves. It is impossible rationally to reconcile claims of traditional authority with needs for continual and radical change. The bowl of sky above our small home has grown to the infinite or near-infinite cosmos; science has decided that the world operates without divine intervention; we ourselves have a material basis and animal ancestry. Religious dictates have wilted under the glare of rationalism; the deity seated on a radiant throne, which was essentially an idealization of the worldly emperor, cannot impress the modern intellectual. Much of existence has become impersonal; the individual understands practically nothing of what countless others have done and are doing to make his world, and society functions more through symbols than real entities in a synthetic environment. Moreover, multiplicity of beliefs in what is essentially a world society tends to neutralize them all; parochial values lose moral force.

Humanity needs a new assessment of the good, with logically convincing bases on which to rest it. But if knowledge destroys old certainties, can it build new ones? Rejection of the idea of obedience to the will of a personal divinity easily leads to rejection of all suprapersonal values. It is more difficult to design a new morality than a space probe.

Science, which remakes our world, should have the compensatory duty of helping us to orient ourselves in it. To some degree, it has done so. Right and justice ought to be based on truth, and science is the systematic search for truth. Its basic honesty and objectivity inspire sounder thinking in all realms, and the selflessness of much science is a model for society. The discoveries of physics, astronomy, and cosmology have given cause for humility and pride, out of which there should come inspiration.

But the scientific temper is partly negative, and conventional science has not been very helpful. Although physicists have become more inclined to talk of something like God in their abstract, non-mechanistic realm, students of behavior would like to exclude everything not clearly measurable and material. The idea that the world is fundamentally simple and accessible to systematic understanding leads directly to an ethics of indifference, in which we are a "chance efflorescence emerging amid the purposeless collapse of the universe."[10]

Despite the enormous and overwhelming growth of complexity and effectiveness, or purposefulness and creativity, in living nature, biologists, following Darwin, are at pains to deny that there can be anything inherently progressive in the process. If a certain primate has attained great power, it is only an accident. As Jacques Monod would have it, "Man knows at last that he is alone in the universe's indifferent immensity, out of which he emerged only by chance."[11] Whether this is true or not has raised far more heated controversy than any other scientific question, perhaps more than all other scientific questions together.

There is no indignity in saying that we and other forms of life have common ancestors. But if we are merely the product of random errors in nucleic acid sequences selected for the ability to survive and reproduce, there is no deeper significance in anything that life may achieve. The reductionist view implies that the purpose of the organism is merely to propagate more of certain sequences of purines and pyrimidines. If the person is only the instrument of genes, why should the instrument care about genes—or anything except pleasing itself? The molecules making up our bodies have no more importance than a wisp of gas in outer space.

Taking the genes too seriously encourages a neo-Darwinist social theory, which in the extreme rationalizes unbridled contestation with different races if not their destruction. In such an approach, the leaders of Germany, which was perhaps the most scientifically-minded nation of the world, claimed in 1933-1945 to be scientific and modern in consecrating collective national and racial selfishness, to the extreme of extermination of "inferiors."

Science has claimed to be above values, which are subjective and inaccessible to objective study. Yet scientists could not go forward without strong and lofty values. Unfortunately, persons deeply dedicated to the expansion of knowledge and to combatting distortions of ignorance and myth, insist, without scientific proof, on the strictly material basis of everything, including ourselves and our minds. It is the business of science to take things apart to find out how they work; this is the only approach that makes sense to most researchers. The scientist naturally inclines to feel that the answer has to be reductionist: all complexity must be derivable from very simple things. The implication is that there are no real values, or that what we take to be values are deceptive—however strongly science itself upholds values.

In the study of the mind and its ways, dissection leads to little enlightenment, even if one has faith that analytic answers to deepest puzzles of the mind lurk hidden somewhere, somehow, in the jumble of billions of neurons. What one might call the profoundest questions, that is, those inaccessible to exact inquiry, are simply ignored in most discussions of functioning of the brain. Biologists likewise have little to say about human affairs. The few who have tried, such as sociobiologists who would base ethics on observations of animal behavior and postulate genes responsible for many aspects of human behavior, have had no success. Animal societies offer many suggestive parallels and may help us understand our failures, but an effort to draw lessons from them for human values misleads because it lowers human motivation to the animal level instead of improving on it. Consideration of animal societies may give some understanding of why we are what we are, but not of what we ought to be or might become; it can teach mostly what has to be transcended.

The gulf between scientific and humanistic cultures is unfortunate. For most scientists the pursuit of facts seems much more promising than "merely" cultural occupations, which lead to no solid results but seem more like recreation and relaxation. On the other hand, humanists seldom can or wish to devote the time and energy needed to digest the culture of science. Neither side has much to say to the other, when they should be joining hands to tackle the questions of our destiny.

It should be the business of professional philosophy to contribute to the guidance of humanity. But the philosophers have mostly failed to compensate for the imbalances of scientific thought and have perhaps disoriented as much as they have oriented. Philosophy usually looks at the world through old spectacles and takes little account of the findings of science, aside from bowing to such unavoidable facts as quantum uncertainty. Although some have taken note of ethical issues, philosophers usually seem to pride themselves on other-worldliness. Dedicated more to the critique of statements than to answering questions, they regard with suspicion anything relevant to the realities of human existence.

The reforging of values for the higher civilization is crucial; we do not choose to have values or not, but what kind we shall have. Appropriate values are quite as vital for civilization as energy resources, and they become the more important as civilization rises in technology, sophistication and artificiality.[12] Most modern problems can be seen as questions of appropriate values.[13] It is difficult to envision a problem, short of an astronomical calamity, not amenable to technology in hand or capable of being devised if the will is strong enough.

## THE GOOD

The mind is neither good nor bad, capable of outrageous baseness as well as saintliness, mostly indifferent. In a sense, there is always an equal amount of good and evil in the world, because we set our standards according to what we know. Yet life can create good, and if ephemeral happiness is for the individual, the good is for the community or the human achievement.

There will always be disagreement regarding matters of deep personal moment. There should be, because enlightened disagreement can lead to more clarity. But right and wrong are decided by the judgment of humanity. Its verdict may be erroneous, in that subsequent judges decide differently; but right and wrong, or good and bad, arise from the needs of the community. Conventions and agreed principles, although often of doubtful rationality, are the indispensable framework of community existence. For all except the insane or extreme eccentrics, social imperatives are compelling; indeed, people uncertain of them feel intensely insecure. There can be no fully rationalistic ethic that we might choose like a suit and wear as we please. There must always be a sense of right and wrong beyond our powers of logical explanation, a given code, which no single individual, however intelligent, can formulate. Good and evil are not invented but learned and understood. Both the serpent in Eden and God declared that to know good and evil is to be godlike.

Yet understanding of good and evil should be related to the meaning of the cosmos and the metacosmos in which it is embedded. Our existence seems to imply that intelligence, creativity, and purpose are values inherent in the nature of things. Most moralists and philosophers of the past would probably agree. Long ago, St. Augustine wrote, "The prime author and mover of the world is intelligence"; he held intelligence and truth to be the end of the universe, the best office of humans being the pursuit of wisdom.[14] The Socratic principle that evil arises from ignorance and good from understanding remains as true as ever.

If we accept kinship with the living world and see ourselves as part of the inscrutable yet meaningful order, our universe seems reasonable and beneficent even though its works are often cruel. The essence of life is directedness; and the more evolved the living form, the more purposeful. In becoming human, an apelike creature laid its stake on the application of intelligence and knowledge to life. Whatever the shortcomings of civilization, there is grandeur in the human component that crowns the living world. In the words of Kallistor Ware, "Such is our condition and opportunity as human beings: we have our feet planted on the ground, but our vision embraces the sky, and it is thus our human privilege to draw the spiritual and the material into unity, to express them as inseparable and complementary aspects of a single and undivided whole."[15]

Civilization is an endless contest between growth and decay, and the world is a little different for each one's passing through. Within the reach of everyone there lies a set of building blocks of various sizes and shapes waiting to be grasped and put in place. From them, no one can make everything to be desired, and sometimes one tries to construct a cathedral upon quicksand or sees it knocked into rubble. But to judge the good one need not add up results. Not all worthy flowers bear seed, and some sterile ones may be better turned than the fertile. All are failures in that they fall short of what they might have been. Virtue is will and intent, not the tally of the harvest. The way or means is superior to the end or result; we are responsible for intentions and actions, but the outcome is subject to chance and the tempests of the world.

The good and the gratifying are usually not far apart, although it is difficult that virtue should be quite happy, because it is forever risking its own happiness. There are always perils and struggles; virtue may merit paradise but cannot enjoy it. It is then the best sorrow to have too little time and strength to live and create.

The more the view turns outward, the less confining are the inevitable limitations of this existence. Perhaps civilization will collapse in fiery death or sink in morbid decay, or go on to greater triumphs of self-compounding intelligence, even conceivably, as some physicists speculate, to the creation of new universes. We do not know—it is better not to know—how long it can go forward and to what heights it can rise.

# 6

# Epilogue

## The unfolding of complexity

Our cosmos, in its incomprehensible immensity, variety, dynamism, and complexity, arose about 15 billion years ago out of a matrix of order and something akin to will, purpose, and intelligence on a higher plane. The cosmos was so shaped that it has attained self-knowledge through us; and we may congratulate ourselves that, in our capacity to understand something of the cosmos and raise the level of order, we are assimilating more of the creative force behind it. We may in the future be replaced by creatures better fulfilling this design in the same degree that we surpass the dinosaurs, for whom, during a hundred million years, it might have seemed that the cosmos was planned. But so far as we know, we are the acme of achievement of 15 billion years.

The cosmos apparently began from something like a dimensionless point, and at the commencement containing no information except its existence. But with expansion, particles and forces differentiated, and the laws of nature came into existence along with the entities to which they pertained. Creation did not cease with the generation of the fireball; structure has been building on itself ever since the beginning as though in fulfillment of a drive to ordered complexity. Quarks segregated into nucleons, which came together with electrons to form atoms. Matter gathered into galaxies and stars with their planets. Complex molecules led to the generation of life, and the responsiveness of living things brought the evolution of intelligence and the emergence of civilization and the mind. A sufficiently structured confluence of material particles became able to approach qualities of the matrix originally giving rise to the particles.

We cannot hope fully to understand this process, just as the computer cannot understand its maker, despite its partial realization of some of the qualities of the higher being. How should we try to make sense of the marvel? A child, taking an old-style alarm clock to pieces, perceives how the pressure of the spring, regulated by the balance-wheel, moves the hands. But reductionism, however attractive as theoretical simplification, is to a large extent only theoretical and programmatic, giving an illusion of deeper understanding than we have. To suppose that chemistry is to be comprehended in terms of quantum mechanics, or biology in terms of Darwinian theory, is very far from actually being able to account for facts in such terms. It is like applying a screwdriver to the guts of a computer.

Even when the more complex ought to be analyzable in terms of simpler entities, mathematical treatment is likely to be impossibly complicated. No one studying petroleum derivatives would consider only the properties of hydrogen and carbon, however useful it is to know that they combine in proper proportions to make oil, gasoline, and so forth. Water is so "creative" that a scientist can spend a lifetime studying its properties. Biologists are happy to learn what they can of the workings of nucleic acid, and molecular biology sheds much light on genetics; but zoologists have no hope that the string of genes, shaped by the mechanics of selection, will answer all their questions. They deal with such matters as reproduction, cooperation, adaptation, instinct, selection, phylogeny; for most of their questions, the details of heredity and physiology are irrelevant, and the survival of the fittest is only a hypothesis in the background.

Sometimes higher-level facts can be satisfactorily explained in lower-level terms. For example, sound and heat are well understood as the results of motions of atoms and molecules. The rainbow becomes understandable through knowledge of the refraction of light by water droplets, and the refraction is explainable in terms of behavior of electrons and photons. Quantum mechanics gives a good description of the basic properties of elements and of matter-radiation interaction. Frequently, on the other hand, study of a complex entity illuminates the behavior of simpler things, as in the laser.[1] The cosmos, in fact, may be built largely by downward causation through selection of random inputs.[2]

To keep within the scope of rational science, we should expand the scope of science, rather than limit the breadth of inquiry. Understanding includes both analysis, or taking apart, and synthesis, or grasp of the whole. The latter comprises description, comparison, and finding meanings and significance in the context of larger questions or learning. Of the two aspects of understanding, finding relations of components and perceiving patterns, the latter is the more important. This is inevitable because of the way the universe is put together, new order coming into being atop simpler components, any whole being composed of simpler things. We do not even feel much need to try to understand many things as long as they fit reasonably well into a broader picture. We cannot fully reduce the properties of ice to those of hydrogen and oxygen, which are largely irrelevant. But it is possible to learn a great deal about ice in the light of knowledge of crystals and solids. The synthetic-holistic approach does not promise the mathematical answers of an analytic-reductionist dissection, but it is necessary for understanding complex entities.

New order grows out of a mixture of uncertainty and regularity, of variation within patterns or stability, as in the turbulence that formed galaxies, the sieve of evolution, and the self-compounding of culture. The freedom of particles to make themselves variably manifest as they "choose" (within narrow limits) opens the way to infinite variety of compounds and combinations. Even if conditions were more exactly known than is possible in view of the obscurity at the basis of all things, uncertainties would rapidly multiply.[3]

The evolution of life has likewise produced unpredictable marvels within the limitations of its material base. As the animal becomes more complex, there is more need for coordination, that is, a message system, becoming ultimately a nervous system to use information from various organs and the outside world. On a higher level, matter is not only guided by physical laws but moved by mental events, as minds react unpredictably to external information and internal drives. When the brain became able to invent and remember useful things, culture became possible—but only when societies were so structured as to reward the making, retention, and spread of knowledge and inventions. Civilization arose out of the better organization of large groups permitting the indefinite accumulation of improvements and compounding the usefulness of the powers of intelligence.

Yet creative intelligence has apparently been only very narrowly victorious; and if the cosmos was designed for intelligence, it was barely so. Intelligence took some 4 billion years to arise on this earth, and high civilization seems not to be common elsewhere in our galaxy. Life is also lucky to have been permitted to continue for so many eons. There have been many catastrophic eliminations of living forms, roughly every 26 million years,[4] which may have been necessary to bring deeper change by destroying a static equilibrium.[5] If the abundance of life has been thus many times gravely set back, it is very good fortune that extinction was never total.

Even in the very latest stages, the triumph of technological civilization has been chancy because of the difficulty of the adjustment of social needs and individual drives. To join individual efforts effectively without losing the freedom that upholds the quality of culture has seldom been well achieved in history. The basic problem in the interrelationship of conflicting natures is to weave the functional web of interacting minds.

At present, the problems and obstacles for further advancement loom more prominently than the means of overcoming them. The chief cause for optimism for the long-term future may be faith in the purpose of the cosmos fulfilling itself. Conceivably the fact that intelligence has succeeded thus far, despite all the hazards, may be evidence of a kindly fate or a predisposing purpose. In order to create intelligent life as we know it, the cosmos had to be too finely tuned to be the result of chance, unless we stretch the imagination to assume an infinity of universes with all possible parameters. It consequently seems reasonable to guess that it was created for the production, or for the expression in our material form of something of the order of the metacosmos; and there is reason to hope that the success of intelligence is more likely than its failure.

Life is impelled to growth, self-perpetuation and expansion, as the tree or fish replaces itself and increases its kind. In a social animal, purpose becomes more diffuse. In the civilized human, it becomes conscious and expands far beyond individual self-preservation and reproduction. For further growth of civilization, even for its long-term continuation, it seems indispensable that humanity adopt the purpose of its collective survival and improvement.

The mind, sharing in the purposes of humanity, seems to join with the metacosmos from which our cosmos was born. It increasingly merges with the matrix of order, intelligence, will, and purpose, for the greater glory of complexity, to ends we cannot imagine.

# Notes

## Preface and Chapter 1

[1] Paul Davies, *The Cosmic Blueprint* (New York: Simon and Schuster, 1988), p.6.

[2] Mary Maxwell, *Human Evolution: A Philosophical Anthropology* (New York: Columbia University Press, 1984), p.2.

[3] See Edward Harrison, *Darkness at Night: A Riddle of the Universe* (Cambridge: Harvard University Press, 1987) pp.196-97.

[4] Joseph Silk,*The Big Bang: The Creation and Evolution of the Universe* (San Francisco: W.H. Freeman, 1980), pp.315-28.

[5] John Gribbin, *In Search of the Big Bang* (New York: Bantam Books, 1986), pp.91-93.

[6] James S. Trefil, *The Moment of Creation*, (New York: Charles Scribner's Sons, 1983); John D. Barrow & Joseph Silk, *The Left Hand of Creation* (San Francisco: W.H. Freeman).

[7] G. Siegfried Kutter, *The Universe and Life: Origins and Evolution* (Boston: Jove and Bartlett, 1987) p.70.

[8] Paul Davies,*God and the New Physics* (New York: Simon & Schuster, 1983), p.42.

[9] Davies, *God and the New Physics*, p.39.

[10] Paul Davies, *The Forces of Nature*, 2nd ed. (New York: Cambridge University Press, 1986), p.167.

[11] John D. Barrow and Frank J. Tipler, *The Anthropic Cosmological Principle* (New York: Oxford University Press, 1986), p.18.

[12] John Gribbin, *In Search of the Big Bang*, p.344.

[13] Paul Davies, *Superforce: The Search for a Grand Unified Theory of Nature* (New York: Simon & Schuster, 1984), p.184.

[14] Davies, *God and the New Physics*, p.179.

[15] *Ibid.*, p.188.

[16] *Ibid.*, p.159.

[17] Paul Davies, *Other Worlds: Space, Superforce, and the Quantum Universe* (New York: Simon & Schuster, 1980), p.88.

[18] Richard Feynman, *QED: The Strange Theory of Light and Matter* (Princeton: Princeton University Press, 1985) p.138.

[19] Gribbin, *In Search of the Big Bang*, p.215.

[20] *Ibid.*, p.257.

[21] Nick Herbert, *Quantum Reality* (Garden City: Doubleday, 1985), p.189.

[22] Gribbin, *In Search of the Big Bang*, p.261.

[23] Heinz R. Pagels, *Perfect Symmetry: The Search for the Beginning of Time* (New York: Simon & Schuster, 1985), pp.300-304.

[24] Lawrence M. Krass, "Dark Matter in the Universe," *Scientific American* 255, December 1986, pp.58-68.

[25] Davies, *Superforce:*, p.148.

[26] A. Zee, *Fearful Symmetry: The Search for Beauty in Modern Physics* (New York: Macmillan, 1986) pp.88, 90.

[27] Terry Goldman, Richard J. Hughes, and Michael M. Nieto, "Gravity and Anti-Matter," *Scientific American* 258, March 1988, pp.48-57.

[28] Davies, *Forces*, p.119.

[29] *Ibid.*, p.164.

[30] Karl R. Popper and John C. Eccles, *The Self and Its Brain* (New York: Springer International, 1977), p.7.

[31] Pagels, *Perfect Symmetry*, p.244.

[32] The theme of Ilya I. Prigogine and Isabelle Stengers, *Order Out of Chaos: Man's New Dialogue with Nature* (New York: Bantam Books, 1984)] passim.

[33] Pagels, *Perfect Symmetry*, p.31.

[34] Rosaria Egidi, "Emergence, Reductionism, and Evolutionary Epistemology," in Gerard Radnitzky and William W. Bartley III, eds., *Evolutionary Epistemology, Rationality, and the Sociology of Knowledge* (La Salle, IL: Open Court Publishing Co., 1987) pp.158-60.

[35] Paul Hoffman, "The Next Leap in Computers," *The New York Times Magazine*, December 7, 1986, p.132.

[36] Cited by Prigogine and Stengers, *Order Out of Chaos*, p.313.

[37] Barrow and Tipler, *The Anthropic Cosmological Principle*, p.127.

## CHAPTER 2

[1] Karl R. Popper, *The Open Universe* (Towata, NJ: Rowman and Littlefield, 1982), p.150.

[2] William Day, *The Search for Life's Beginnings* (New Haven: Yale University Press, 1984), pp.2-4.

[3] John Gribbin, *Genesis; The Origins of Man and the Universe* (New York: Delacorte, 1981), p.183.

[4] Graham C. Smith, "The Clay Life Hypothesis," *The World and I*, December 1987, p.181.

[5] William Day, *The Search* (New Haven: Yale University Press, 1984), p. 131.

[6] G. Siegfried Kutter, *The Universe and Life: Origins and Evolution* (Boston: Jove and Bartlett, 1987) pp.326-28.

[7] *Ibid.*, p.23.

[8] As suggested by the "Gaia hypothesis" of James Lovelock.

[9] Kutter, *The Universe and Life*, p.260.

[10] Charles J. Lumsden & Edward O. Wilson, *Genes, Mind and Culture: The Coevolutionary Process* (Cambridge: Harvard University Press, 1983), p.697.

[11] Salvador E. Luria, Stephen Jay Gould, Sam Jinjer, *A View of Life* (Menlo Park: Benjamin/Cummings Publishing Co., 1981), pp.651-53.

[12] Mary Midgley, *Beast and Man: The Roots of Human Nature*, (Ithaca: Cornell University Press, 1978), p.153.

[13] Francis Hitching, *The Neck of the Giraffe* (Garden City, NY: Anchor Books, 1978) p.166.

[14] Mark A. McMenamin, "The Emergence of Animals," *Scientific American*, 256, April 1987, p.94.

[15] Pierre Paul Grassé, *The Evolution of Living Organisms* (New York: Academic Press, 1978), pp. 82-84.

[16] Anthony Smith, *The Mind* (Hammondsworth: Penguin Books, 1984)

[17] Robert Ardrey, *The Social Contract* (New York: Atheneum, 1970), p.82.

[18] *Ibid.*, pp.61-62.

[19] Bernard Campbell, *Human Evolution: An Introduction to Man's Adaptations* 3rd ed. (New York: Aldine Publishing Co., 1985), passim.

[20] Michael Denton, *Evolution: A Theory in Crisis* (Bethesda MD: Adler and Adler, 1986) p. 108.

[21] Verne Grant, *The Evolutionary Process, A Critical Review of Evolutionary Theory* (New York, Columbia University Press, 1985), pp.10-11.

[22] Theodosius Dobzhansky and Ernest Boesiger, *Human Culture: A Moment in Evolution* (New York: Columbia University Press, 1983) p.2.

[23] Douglas J.Futuyma, *Evolutionary Biology*, 2nd ed. (Sunderland MA: Sinauer Associates, 1986) p.7.

[24] Denton, *Evolution*, pp.109-10, 164.

[25] Denton, *Evolution*, p.193.

[26] Denton, *Evolution*, pp.211-12.

[27] Grassé, *Evolution*, pp.56-57.

[28] Peter R. Grant, *Evolution and Ecology of Darwin's Finches* (Princeton: Princeton University Press, 1986), pp.401, 403.

[29] Lubert Stryer, "The Molecules of Visual Excitation," *Scientific American* 257, July 1987, pp. 42-51.

[30] Grassé, *Evolution*, p.136.

[31] Milton Love, "The Alien Strategy," *Natural History* 89, May 1980, p.30.

[32] Love, "Alien Strategy," p.31.

[33] D.W.T. Crompton and S.M. Joyne, *Parasitic Worms* (New York: Crane, Russak & Co., 1980), p.30.

[34] Peter R. Grant, *Evolution and Ecology of Darwin's Finches* (Princeton: Princeton University Press, 1986), pp. 372, 374.

[35] Bernhard Grzimek, *Grzimek's Animal Life Encyclopedia*, vol. 2 (New York: Van Nostrand, 1975), p.184.

[36] Richard Conniff, "The Little Suckers Have Made a Comeback," *Discover* 8, August 1987, p.88.

[37] Andrew Cockburn and Anthony K. Lee, "Marsupial Femmes Fatales," *Natural History* 97, March 1988, p.43.

[38] A.F. Dixon, *The Natural History of the Gorilla* (New York: Columbia University Press, 1981), p.54.

[39] Spencer C. Barrett, "Mimicry in Plants," *Scientific American*, 257, September 1987, p.78.

[40] Ernest Mayr, *Evolution and the Diversity of Life* (Cambridge: Harvard University Press, 1976), pp.88-111.

[41] *Science News*, May 30,1987, p.340.

[42] Guenter Waechterhaeuser, "Light and Life: On the Nutritional Origins of Sense Perception," in Gerard Radnitzky and William W. Bartley, eds., *Evolutionary Epistemology, Rationality, and the Sociology of Knowledge* (La Salle, IL: Open Court Publishing Co., 1987), pp.121-39.

[43] Karl R. Popper & John C. Eccles, *The Self and its Brain* (New York: Springer International, 1977), p.12.

[44] Grant, *Evolutionary Process*, pp. 182-83.

[45] Hitching, *The Neck*, p.147.

[46] *Science News*, 131, April 4, 1987, p.214.

[47] Rupert Sheldrake, *A New Science of Life: HQ 331 S 439 Biology* (Los Angeles: J.P. Tarcher Inc., 1981), p.20.

[48] Grassé. *Evolution*, p.166.

[49] Arthur Koestler, *The Ghost in the Machine* (New York: Macmillian, 1968), p.197.

[50] Robert Rosen, "Organisms as Causal Systems," in *Theoretical Biology and Complexity* (New York: Academic Press, 1985), ed. Robert Rosen, pp.200, 202.

## CHAPTER 3

[1] Bernd Würsig, "The Behavior of Baleen Whales," *Scientific American* 259, April 1988, p.107.

[2] Verne Grant, *The Evolutionary Process, A Critical Review of Evolutionary Theory* (New York: Columbia University Press, 1985) pp.356-57.

[3] Jane Goodall, *The Chimpanzees of Gombe: Patterns of Behavior* (Cambridge: Harvard University Press, 1986), pp.28-30, 34, 133-36.

[4] *Ibid.*, pp.561-63.

[5] *Ibid.*, p.351.

[6] Michael Ghiglieri, "War Among the Chimps," *Discover* 8, November 1987, pp.67-76.

[7] Robert Ardrey, *The Social Contract* (New York: Atheneum, 1970), p.167.

[8] Curtis G. Smith, *Ancestral Voices: Language and the Evolution of Human Communication* (Englewood Cliffs: Prentice Hall, 1985), p.15.

[9] Mary Maxwell, *Human Evolution: A Philosophical Authropology* (New York: Columbia University Press, 1984), p.197.

[10] Bernard Campbell, *Human Evolution: An Introduction to Man's Adaptations*, 3rd ed. (New York: Aldeine Publishing Co., 1985), p.290.

[11] Charles J. Lumsden & Edward O. Wilson, *Genes, Mind, and Culture: The Coevolutionary Process* (Cambridge: Harvard University Press, 1981), p.327.

[12] Bernard Campbell, *Human Evolution*, p.280.

[13] Smith, *Ancestral Voices:*, p.29.

[14] Wilson, *Sociobiology* (Englewood Cliff: Prentice Hall, 1985), p.271.

[15] Campbell, *Human Evolution*, p.317.

[16] Grant, *The Evolutionary Process*, p.421.

[17] Maxwell, *Human Evolution*, p. 195.

[18] Campbell, *Human Evolution*, p.282.

[19] Anthony Smith, *The Mind* (Hammondsworth: Penguin Books, 1984), p.4.

[20] *Ibid.*, pp.14-15.

[21] Curtis G. Smith, *Ancestral Voices*, p.29.

[22] *Ibid.*, pp.41, 24, 36.

[23] John N. Wilford, "Artistry of the Ice Age," *The New York Times Magazine*, October 12, 1986, p.47.

[24] Robin Dennell, "Needles and Spear-throwers," *Natural History* 95, October 1986, pp.76-79.

[25] Robert Wesson, *The Imperial Order* (Berkeley: University of California Press, 1967).

[26] Robert Wesson, *State Systems: International Pluralism, Politics, and Culture* (New York: Macmillan, 1978).

[27] Philip J. Davies & Ruben Hersh, *The Mathematical Experience* (Boston: Houghton-Miffin, 1981), p. 34.

[28] Heinz R. Pagels, *Perfect Symmetry The Search for the Beginning of Time* (New York: Simon & Schuster, 1985) p.50.

[29] Edward R. Harrison, *Cosmology: The Science of the Universe* (New York: Cambridge University Press, 198l), p. 398; G. Siegfried Kutter, *The Universe and Life: Origins and Evolution* (Boston: Jove and Bartlett, 1987), pp.357-65.

[30] Bruce Bower, "IQ's Generation Gap," *Science News* 132, August 15, 1987, p.108.

[31] John D. Barrow & Frank J. Tipler, *The Anthropic Cosmological Principle* (New York: Oxford University Press, 1986), p.133.

## CHAPTER 4

[1] *Business Week*, June 2, 1986, p.93

[2] Karl R. Popper, *The Open Universe* (Towata, NJ: Rowman and Littlefield, 1982), p.151.

[3] Howard Gardner, *The Mind's New Science: A History of the Cognitive Revolution* (New York: Basic Books, 1985) p.383.

[4] Paul Davies, *Superforce: The Search for a Grand Unified Theory of Nature* (New York: Simon & Schuster, 1984), p.49; Paul Davies, *Other*

*Worlds: Space, Superforce, and the Quantum Universe* (New York: Simon & Schuster, 1980), p.123.

[5] C.F. von Weizsäcker, *Science News* 132, July 11, 1987, p.27.

[6] John Searle, *Minds, Brains, and Science* (Cambridge: Harvard University Press, 1984), p.13.

[7] Roger W. Sperry, *Science and Mind* (New York: Columbia University Press, 1983), p.85.

[8] Mario Bunge, *The Mind-Body Problem, A Philosophical Approach* (New York: Press, 1980), p.16.

[9] Howard Gardner, *The Mind's New Science* (New York: Basic Books, 1985), p.283.

[10] Curtis G. Smith, *Ancestral Voices: Language and the Evolution of Human Communication* (Englewood Cliff: Prentice Hall, 1985), p.113.

[11] Wilder Penfield, *The Mystery of the Mind* (Princeton: Princeton University Press, 1975).

[12] Ludwig van Bertalanffy, *A Systems View of Man* (Boulder, CO: Westview, 1981) p.80.

[13] *Ibid.*, p.89.

[14] Edward O. Wilson, *On Human Nature* (Cambridge: Harvard University Press, 1978), p.73.

[15] Roger W. Sperry, "Bridging Science and Values," in John Eccles ed., *Mind and Brain: The Many Faceted Problem* (Washington: Paragon House, 1982), p.260.

[16] Paul Davies, *The Cosmic Blueprint* (New York: Simon and Schuster, 1988) p.174.

[17] Kenneth Denbigh, *An Inventive Universe* (New York: George Braziller, 1975) p.175.

[18] Gordon R. Taylor, *The Natural History of the Mind* (New York: E.P. Dutton, 1979), p.41.

[19] Charles J. Lumsden & Edward O. Wilson, *Genes, Mind, and Culture: The Coevolutionary Process* (Cambridge: Harvard University Press, 1981), p.328.

[20] Howard Gardner, *Frames of Mind: The Theory of Multiple Intelligence* (New York: Basic Books, 1985), p.86.

[21] Judith Hooper and Dick Teresi, *The Three-Pound Universe* (New York: Dell, 1987), passim.

[22] Taylor, *Natural History*, p.96.

[23] *Ibid.*, p.107.

[24] Michael S. Gazzaniga, *The Social Brain: Discovering the Networks of the Mind* (New York: Basic Books, 1985), p.86.

[25] Gardner, *Frames of Mind.*

[26] Michael S. Gazzaniga, *The Social Brain: Discovering the Networks of the Mind* (New York: Basic Books, 1985), Mario Bunge, *The Mind-Body Problem, A Psychological Approach,* p.190; Sally P. Springer and Gery Deutsch, *Left Brain, Right Brain* (San Francisco: W.H. Freeman, 1981), p.183.

[27] Wilder Penfield, *The Mystery of the Mind* (Princeton: Princeton University Press, 1975), p.55.

[28] *Ibid.,* pp.76-78.

[29] Derek Parfit, "Divided Minds and the Nature of Persons," in Colin Blakemore and Susan Greenfield eds. *Mindwaves: Thoughts on Itelligence, Identity, and Consciousness* (Oxford: Basil Blackwell, 1987), p.25.

[30] Bunge, *The Mind-Body Problem,* p.6.

[31] Hooper and Teresi, *Three-Pound Universe,* pp.135-37.

[32] Taylor, *Natural History,* p.43.

[33] Calder, *The Mind of Man* (New York: Viking Press, 1970), p.88.

[34] Hooper and Teresi, *Three-Pound Universe,* pp.90-94.

[35] Roger W. Sperry, *Science and Moral Priorities* (New York: Columbia University Press, 1983), p.40.

[36] John C. Eccles, "Cerebral Activity and Freedom of the Will," in Eccles ed. (Washington: Paragon House, 1982), p.151.

[37] Gordon R. Taylor, *Natural History,* p.115.

[38] Paul Davies, *God and the New Physics* (New York: Simon & Schuster, 1983), p.140.

[39] Hooper and Teresi, *Three-Pound Universe,* p.366.

[40] Nick Herbert, *Quantum Reality* (Garden City: Doubleday, 1985).

[41] Gordon R. Taylor, *Natural History,* p.136.

[42] *Science News* 130, November 15, 1986, p.130.

[43] James Goodwin and Jeane Goodwin, "Impossibility in Medicine," in Davis and Park eds., *No Way: The Nature of the Impossible* (New York: W.H. Freedman, 1987) p.55.

[44] Gardner, *Frames of Mind:,* p.42.

[45] Taylor, *Natural History,* pp.139-140.

[46] *Ibid.,* pp.5, 101.

[47] Kenneth R.Pelletier, *A New Age, Problems and Potential* (San Francisco: Robert Briggs Associates, 1985), p.36.

[48] Gardner, *Frames of Mind*, p.86.

[49] Nigel Calder, *The Mind of Man* (New York: Viking Press, 1970), pp.86-87.

[50] Barbara B. Brown, *Supermind: The Ultimate Energy* (New York: Harper & Row, 1980), p.47.

[51] Hooper and Teresi, pp.306-7.

[52] Hooper and Teresi, pp.61, 62.

[53] Popper, *Open Universe*, p.157; Gardner, *The Mind's New Science*, p.277.

[54] Eric P. Ploten, *Critique of the Psycho-Physical Identity Theory* (The Hague: Mouton, 1973), p.55.

[55] von Bertalanffy, *A Systems View of Man*, p.91.

[56] Janos Szentagathai, "The Brain-Mind Relation: A Pseudoproblem," in Blakemore and Greenfield eds., *Mindwaves*, p.324.

[57] Karl Popper, "Natural Selections and the Emergence of Mind," in Gerard Radnitzky and William W. Bartley, III, eds., *Evolutionary Epistemology, Rationality, and the Sociology of Knowledge* (La Salle, IL: Open Court Publishing Co., 1987), p.152.

[58] Brown, *Supermind*.

[59] Penfield, *The Mystery of the Mind*, p.47.

[60] Taylor, *Natural History*, p.84.

[61] Brown, *Supermind*, p.123.

[62] von Bertalanffy, *A Systems View*, p.93.

[63] Sally P. Springer and Gery Deutsch, *Left Brain, Right Brain* (San Francisco: W.H. Freeman, 1981), pp.184, 194.

[64] von Bertalanffy, *A Systems View*, p.96.

[65] Smith, *Ancestral Voices*, p.129.

[66] Noam Chomsky, *Language and Mind* (New York: Harcourt Brace Jovanovich, 1972).

[67] Bertalanffy, *A Systems View*, pp.956.

[68] *Ibid.*, p.94.

[69] Mario Bunge, *The Mind-Body Problem*, p.52.

[70] Davies, *God and the New Physics*, p.724.

[71] Philip J. Davis and Ruben Hersh, *The Mathematical Experience* (Boston: Houghton-Miffin, 1981), pp.401, 404.

[72] Davies, *God and the New Physics*, p.149.

[73] Richard Courant and Herbert Robbins, *What is Mathematics?* (New York: Oxford University Press, 1953), p.217.

[74] Heinz R. Pagels, *Perfect Symmetry: The Search for the Beginning of Time* (New York: Simon & Schuster, 1985), p.265.

[75] Gribbin, *In Search of the Big Bang*, p.307.

[76] Saunders MacLane, *Mathematics, Form and Function* (New York: Springer Verlag 1986), p.146.

[77] Daniel Gorenstein, "The Enormous Theorem," *Scientific American* 253, December 1985, pp.104-115.

[78] Joe P. Buhler, "Of Primes and Pennies," *Science News* 85, November, p.84.

[79] Morris Kline, *Mathematics: The Loss of Certainty* (New York: Oxford University Press, 1980) pp. 322-26.

[80] Kline, *Mathematics*, p.314.

[81] Davis and Hersh, *The Mathematical Experience*, p.334.

[82] Edward R. Harrison, *Masks of the Universe* (New York: Macmillan Publishing Co.,1985),

[83] Howard W. Eves and Carroll V. Newson, *An Introduction to the Foundation and Fundamental Concepts of Mathematics* (New York: Holt Reinhart & Winston, 1965), p.336.

[84] Kline, *Mathematics*, p.263.

[85] Ivars Peterson, "Escape into Chaos," *Science News* 125, May 26, 1984, p.328.

[86] Philip J. Davis and Ruben Hersh, *The Mathematical Experience*, p.87.

## CHAPTER 5

[1] May Pines, "Is Sociobiology All Wet?", *Psychology Today*, (11), May 1978, p.23.

[2] Edward Wilson, *Sociobiology* (Cambridge: Harvard University Press, 1980), p.3.

[3] *Ibid.*, p.286.

[4] Theodosius Dobzhansky and Ernest Boesiger, *Human Culture: A Moment in Evolution* (New York: Columbia University Press, 1983), p.73.

[5] Wilson, *Sociobiology*, Ch.5.

[6] Gordon R. Taylor, *The Natural History of the Mind* (New York: E.P.Dutton,1979), p.60.

[7] Aleksandr Solzhenitsyn, *The Gulag Archipelago* (New York: Harper & Row, 1976), vol. 2, p. 490.

[8] Abraham H. Maslow, *The Farther Reaches of Human Nature* (New York: Macmillan, 1971), p.45.

[9] Mario Bunge, *The Mind-Body Problem, A Psychological Approach* (New York: Pergamon, 1980), p.207.

[10] Peter Atkins, "Purposeless People," in Arthur Peacocke and Grant Gillett eds., *Person and Personality: A Contemporary Inquiry* (Oxford: Basil Blackwell, 1987), p.23.

[11] Cited by Prigogine and Stengers, *Order Out of Chaos: Man's New Dialogue with Nature* (New York: Bantam Books), p.22.

[12] Roger W. Sperry, "Bridging Science and Values," John Eccles ed., *Mind and Brain: The Many Faceted Problem* (Washington: Paragon House, 1982), p.256.

[13] Roger W. Sperry, *Science and Moral Priority* (New York: Columbia University Press, 1983), p.5.

[14] Howard Gardner, *Frames of Mind: The Theory of Multiple Intelligences* (New York: Basic Books, 1985), p.6.

[15] Kallistor Ware, "The Utility of the Person according to the Greek Fathers," in *Person and Personality*, ed. Peacocke and Gillett, p.202.

## CHAPTER 6

[1] Ludwig van Bertalanffy, *A Systems View of Man* (Boulder: Westview, 1981), p.89.

[2] Edward O. Wilson, *On Human Nature* (Cambridge: Harvard University Press, 1978), p.73.

[3] Paul Davies, *The Cosmic Blueprint* (New York: Simon & Schuster, 1988), p.30.

[4] Verne Grant, *The Evolutionary Process, A Critical Review of Evolutionary Theory* (New York: Columbia University Press, 1985), p.463.

[5] Davies, *The Cosmic Blueprint*, p.114.

# Index

V



W